NATIONAL DEFENSE RESEARCH INSTITUTE

T0108756

Understanding Government Telework

An Examination of Research Literature and Practices from Government Agencies

Cortney Weinbaum, Bonnie L. Triezenberg,
Erika Meza, David Luckey

Prepared for the National Geospatial-Intelligence Agency

For more information on this publication, visit www.rand.org/t/RR2023

Library of Congress Cataloging-in-Publication Data is available for this publication.
ISBN: 978-1-9774-0047-5

Published by the RAND Corporation, Santa Monica, Calif.
© Copyright 2018 RAND Corporation
RAND® is a registered trademark.

Cover photo: AleksandarNakic, Getty Images.

www.rand.org

Preface

The RAND National Defense Research Institute assisted the National Geospatial-Intelligence Agency (NGA) in understanding how its workforce can work in unclassified environments, including outside Sensitive Compartmented Information Facilities (SCIFs). This is the first of two reports from the project, and it examines literature about telework and telework practices from across government agencies. This report will be of interest to government leaders who are currently implementing telework or considering implementing telework. The report identifies technological, legal, policy, financial, and security considerations and examines lessons learned from seven different agencies. RAND selected federal agencies that had telework programs in place, agencies whose employees handle sensitive or classified information, and agencies with data publicly available about these programs. These agencies were the Federal Emergency Management Agency, General Services Administration, Internal Revenue Service, National Aeronautics and Space Administration (NASA), National Science Foundation, Nuclear Regulatory Commission, and United States Patent and Trademark Office. This research and analysis was conducted between April 2016 and April 2017.

A companion report, *Moving to the Unclassified: How the Intelligence Community Can Work from Unclassified Facilities*, draws on the results of this report to inform intelligence agencies about what to consider before conducting work outside secure government facilities.

This research was sponsored by the Human Development Directorate at NGA and conducted within the Cyber and Intelligence Policy Center of the RAND National Defense Research Institute, a federally funded research and development center sponsored by the Office of the Secretary of Defense, the Joint Staff, the Unified Combatant Commands, the Navy, the Marine Corps, the defense agencies, and the defense Intelligence Community. RAND partnered with the Maxwell School of Citizenship and Public Affairs at Syracuse University in conducting the research for this report.

For more information on the RAND Cyber and Intelligence Policy Center, see www.rand.org/nsrd/ndri/centers/intel or contact the director (contact information is provided on the webpage).

Contents

Figure and Tables

Figure

Tables

Summary

Across the federal government, telework is the principal method for allowing employees to work outside agency facilities. This report provides an overview of the literature on telework and explains how government agencies benefit when employees engage in telework. In national security agencies, the benefits of working outside government facilities must be balanced with the need to protect classified and sensitive information. This includes requirements to protect information technology systems from external threats and capabilities to monitor employees for insider-threat risks. Across government, millennials entering the workforce have begun to challenge the status quo with their unwillingness to accept traditional office arrangements, leading some agencies to consider telework and other flexible workplace practices. Such workforce changes have disrupted long-standing recruiting, hiring, education, and workplace practices and increased the demand for telework policies and processes.

This report examines previous studies and surveys related to telework, as well as telework practices at seven federal agencies. We selected agencies that had telework programs in place, agencies whose employees handle sensitive or classified information, and agencies with data publicly available about their programs. These agencies were the Federal Emergency Management Agency, General Services Administration, Internal Revenue Service, National Aeronautics and Space Administration (NASA), National Science Foundation, Nuclear Regulatory Commission, and United States Patent and Trademark Office.

Among the federal programs examined, we found similarities across successful agency telework programs that allowed us to categorize important factors, including compliance with federal and organizational policies, technological accommodations for employees, a measurable return on investment, the adaptation of performance management tools, and training.

For federal agencies interested in implementing a new telework program or modifying an existing one, we found that a clear understanding of the purposes of such a program is essential to guiding the development of program goals, policies, and performance measures, as well as for the managers who will be responsible for developing and implementing new technology capabilities, security protocols, and training. We recommend the following actions for federal leaders implementing a telework program:

- **Establish program goals that clearly explain the mission value of telework** and effectively communicate those goals to the workforce. The agencies we examined set goals to reduce real estate costs, improve employee job satisfaction, and be more responsive to the public and during crisis events.
- **Clearly communicate which job positions are eligible for telework** and which functions within each job position are suitable for off-site work. If the agency has sensitive data that require special handling, employees should be informed and trained on how to work remotely and what security protocols are required. Establishing clear policies and providing adequate training are essential to implementing the parameters that agency leaders set.
- **Create policies that document the agency's implementation of telework,** how data should be handled, and the use of government and personal computing devices. Employees and managers should have a clear understanding of whether telework is acceptable at the agency, how to effectively engage in telework, and what is expected of the teleworking employee and the teleworker's supervisor.
- **Create performance measures for the agency and teleworkers.** Agencies should measure the performance of the telework program against the established goals. For employees and managers, performance measures may consider deliverable-based or

results-oriented management approaches or quantifiable metrics for performance.

Federal agencies with telework programs may benefit from sharing best practices and lessons learned to identify opportunities for improvement and enhancement of their programs. Agencies without telework programs may also benefit from this by informing their discussions about whether a telework program is appropriate for their missions. This report can serve as a tool in understanding mechanisms that can be used to accommodate changing workforces that demand flexible work hours and the option to work from alternate locations.

Acknowledgments

We partnered with Syracuse University's Maxwell School of Citizenship and Public Affairs in conducting the research for this report. Under the guidance of VADM (ret.) Robert Murrett (a former director of the National Geospatial-Intelligence Agency), graduate students Jasmeen Braich, Ariel Gould, Caitlin Hoover, Patricia Latendresse, and Layvon Washington identified federal agencies with effective telework programs, conducted open-source research on each agency, and conducted interviews with selected government officials. This report would not have been possible without their contributions and Professor Murrett's leadership and assistance.

We thank Sina Beaghley, associate director of the RAND Cyber and Intelligence Policy Center, and Katharine Webb, adjunct policy researcher at RAND, for their careful reviews of this report and the feedback they provided.

Abbreviations

COOP	continuity of operations
FEMA	Federal Emergency Management Agency
FY	fiscal year
GAO	U.S. Government Accountability Office
GSA	General Services Administration
IC	Intelligence Community
IRS	Internal Revenue Service
IT	information technology
NASA	National Aeronautics and Space Administration
NGA	National Geospatial-Intelligence Agency
NRC	Nuclear Regulatory Commission
NSF	National Science Foundation
OIG	Office of Inspector General
OPM	Office of Personnel Management
PII	personally identifiable information
ROI	return on investment
SBU	Sensitive But Unclassified

SCIF	Sensitive Compartmented Information Facility
TEA	Telework Enhancement Act of 2010
TIGTA	Treasury Inspector General for Tax Administration
USPTO	United States Patent and Trademark Office
VPN	virtual private network

CHAPTER ONE

Introduction

Agency leaders at the National Geospatial-Intelligence Agency (NGA), a member of the Intelligence Community (IC), have identified a need to support a geographically dispersed workforce. NGA creates unclassified intelligence and collaborates with industry and academia, and its workforce requires solutions for working outside Sensitive Compartmented Information Facilities (SCIFs) to conduct these activities. NGA intelligence officers use unclassified satellite imagery and other types of data to create unclassified intelligence for users who neither need nor have access to secure networks. Moving these employees from SCIFs to unclassified facilities would allow them to work with the data and on the networks that are most effective for their roles, rather than working on the classified information technology (IT) networks traditionally used for intelligence activities. However, this change brings some challenges, since intelligence agencies, including NGA, built their business functions—including email, personnel systems, and time and attendance systems—on classified IT systems that are not accessible outside SCIFs.

Government agencies benefit when employees work outside agency facilities. Alternative locations range from near industry, to academic institutions, to homes. Alternate localities may improve collaboration across sectors, provide a work-life benefit to staff who do not want to commute or relocate, and allow employees to respond more quickly during crisis events when government buildings are not accessible or open. As millennials have entered the workforce,[1] they have begun to challenge the status quo in IC workplaces by disrupting long-standing recruiting, hiring, education, and workplace practices. Compared with previous generations, millennials demand more-flexible work hours, greater ability to work from any location, and access to technology not currently allowed in SCIFs. Millennials are willing to quit jobs or decline job offers that do not provide these benefits, sometimes at the cost of a reduction in salary.[2] Across the federal government, telework is the principal method for allowing work outside agency facilities. Telework drives the creation of policies and processes, the design and implementation of training, and the development of tools needed to facilitate these activities.

National security agencies interested in exploring options for working remotely must balance the need to secure and protect classified and sensitive information with their desire to recruit and retain a millennial workforce. To assist NGA and other agencies striving to conduct more work outside secure government facilities, we conducted a literature review of previous research and studies related to telework, telecommuting, and alternative work locations. We then examined seven federal agencies that have implemented telework to identify lessons learned. A team of graduate students at the Maxwell School of Citizenship and Public Affairs at Syracuse University assisted us by identifying agencies that had telework programs in place, whose employees handle sensitive or classified information, and that provide publicly available data about their programs. With the team from Maxwell, we analyzed data the agencies provided to the Office of Personnel Management (OPM) about the status of their programs, employee manuals and other similar information the agencies publish about their telework programs, U.S. Government Accountability Office (GAO) reports, and other documents. The Maxwell students conducted interviews

---

[1] *Millennials* are broadly defined as individuals born between 1980 and 2004.

[2] For additional research and analysis on the implications of millennials on intelligence, see Cortney Weinbaum, Richard Girven, and Jenny Oberholtzer, *The Millennial Generation: Implications for the Intelligence and Policy Communities*, Santa Monica, Calif.: RAND Corporation, RR-1306-OSD, 2016.

with officials from the telework programs to fill in gaps where publicly available information was not available. This research and analysis was conducted between April 2016 and April 2017.

This, the first of two related reports, examines telework across sectors and in government by reviewing existing studies and examining practices used. The second report, *Moving to the Unclassified: How the Intelligence Community Can Work from Unclassified Facilities*,[3] uses the research in this report to make recommendations for IC agencies seeking to move more of their functions outside SCIFs.

In this report, Chapter Two presents the results of our review of literature on costs and benefits of telework, Chapter Three presents telework practices at seven government agencies, and Chapter Four provides conclusions and recommendations for the federal leaders seeking next steps for their own agencies.

---

[3] Cortney Weinbaum, Arthur Chan, Karlyn D. Stanley, and Abby Schendt, *Moving to the Unclassified: How the Intelligence Community Can Work from Unclassified Facilities*, Santa Monica, Calif.: RAND Corporation, RR-2024-OSD, 2018.

CHAPTER TWO

Literature Review of Telework

We reviewed studies on telework and telecommuting to examine research on best practices, lessons learned, measures of effectiveness, investment costs, and return on investment (ROI) for transitioning an organization into a flexible workplace. We sought studies on the transition to a flexible workplace and how the transition affects recruiting, retention, training, onboarding, career development and promotions, technology, and infrastructure. Given the rapid evolution of the telecommunications technologies that shape telework, we gave precedence in our literature search to studies written within the past decade; however, we also included foundational studies as a baseline.

In our literature review, we first sought to understand current definitions of *telework* and *telecommuting*. We reviewed existing studies on telework to discover benefits, costs, and enablers to consider when formulating strategies, plans, policies, and practices for telework. Next, we examined estimates of the numbers of employees who telework. Then we looked at cost-benefit studies and organized this section of our literature review based on a framework first proposed by Jack Nilles and colleagues in 1976 that examines telework costs and benefits at three levels: the employee, the employer, and the environment.[1] Finally, we sought literature that examined the organizational and technical enablers of telework.

---

[1] Jack M. Nilles, F. Roy Carlson, Jr., Paul Gray, and Gerhard J. Hanneman, *The Telecommunications-Transportation Tradeoff: Options for Tomorrow*, New York: John Wiley and Sons, 1976.

Definition of *Telework*

There are various contradictory definitions of *telework*, partly because many dictionaries, including Merriam-Webster, define *telecommute* but not *telework*. Merriam-Webster defines *telecommute* as, "to work at home by the use of an electronic linkup with a central office."[2] The Telework Enhancement Act of 2010 (TEA) defines *telework* as "a work flexibility arrangement under which an employee performs the duties and responsibilities of such employee's position, and other authorized activities, from an approved work site other than the location from which the employee would otherwise work."[3] Government telework provides employees the opportunity to work from approved alternate locations, such as their homes, separate work sites, or telework centers.

Some research distinguishes between telecommuting and tele-working based on whether working from home actually reduces commuting.[4] For these researchers, employees who do not commute to the office at all during the day are said to telecommute, while those who work from home after returning from the office are said to telework. However, other researchers reserve the word *telework* to refer to the subset of telecommuting that is based at home rather than based in a remote office, and still others exclude those who work from home outside normal business hours from the definition of *telework*.[5]

Some researchers include mobile workers—employees in sales, home repair, truck driving, and so on whose assigned work includes travel to customer or remote locations—as teleworkers, but most researchers exclude mobile workers.[6] Many researchers also exclude the

---

[2]   The definition can be found on Merriam-Webster's website, www.merriam-webster.com (as of February 14, 2017).

[3]   Public Law 111-292, The Telework Enhancement Act of 2010, December 9, 2010.

[4]   Nilles et al., 1976.

[5]   Patricia L. Mokhtarian, "Defining Telecommuting," *Transportation Research Record*, Vol. 1305, 1991; Tammy D. Allen, Timothy D. Golden, and Kristen M. Shockley, "How Effective Is Telecommuting? Assessing the Status of Our Scientific Findings," *Psychological Science in the Public Interest*, Vol. 16, No. 2, 2015.

[6]   See, for example, Patricia L. Mokhtarian and Ilan Salomon, "Modeling the Desire to Telecommute: The Importance of Attitudinal Factors in Behavioral Models," *Transportation*

self-employed who work from home.[7] The distinction of who is self-employed is becoming increasingly blurred as more firms use independent contractors to perform work from home or another location that an employee would otherwise do from an employer-assigned workplace.

Given these contradictory definitions, we have adopted the following definition of *telework*: *All assigned work activities performed outside a person's assigned place of work.* Our definition is independent of whether that work takes place at home, at the local café, or at some other work center; the definition is also independent of whether the work is in addition to or in lieu of travel to the workplace. Our definition excludes workers whose assigned place of work is in the field. This definition is not the same as the one used by the federal government. We were unable to use the government's definition during our literature review, because most studies do not specify where the telework occurs, only that it occurs somewhere other than the employer's office. The government requirement for employees to have an "approved work site" limits flexibility to workers who want to work from anywhere.

Expansion of Telework

In January 2016, the Global Workplace Analytics website estimated that between 20 and 25 percent of workers telework on a frequent basis.[8] The Bureau of Labor Statistics found that 24 percent of employed people did some or all of their work at home in 2015.[9] Within the federal government, OPM reported that, "from 2013 to 2015, telework participa-

---

Research Part A: Policy and Practice, Vol. 31, No. 1, 1997.

[7] Alina Tugend, "It's Unclearly Defined, but Telecommuting Is Fast on the Rise," *New York Times*, March 7, 2014.

[8] Kate Lister, "Latest Telecommuting Statistics," webpage, GlobalWorkplaceAnalytics.com, January 2016.

[9] Bureau of Labor Statistics, U.S. Department of Labor, "24 Percent of Employed People Did Some or All of Their Work at Home in 2015," July 8, 2016.

tion increased from 39 percent to 46 percent of eligible employees and from 17 percent to 20 percent of all employees."[10]

In estimating the number and frequency of teleworkers, some counts include all workers who occasionally work outside the office,[11] while others include only those who regularly work from home or a telework center.[12] For those estimates that include both occasional and regular teleworkers, telework outside normal office hours, such as on evenings or weekends, appears to have remained about the same or to have slightly declined since 2000, while telework during the regularly scheduled workday has increased.[13] It is difficult to determine whether this is because of greater mobility—workers taking time at a coffee shop during the workday to write a report or answer urgent emails on their smartphones, for example—or whether it is the result of workers shifting to home offices.

For a given definition and measurement method, it should be feasible to predict trends in telework; however, this has not proven to be true in practice. Global Workplace Analytics found that 80 to 90 percent of the U.S. workforce has expressed a desire to telework part time, the equivalent of two to three days per week, to balance their independent and collaborative work responsibilities.[14] On the supply side, easily portable information work continues to displace manufacturing and other services that require employees to be physically present in the workplace.[15] Computers and other technology, communication speed and access, and virtual collaboration tools continue to expand

---

[10] OPM, *Status of Telework in the Federal Government: Report to Congress; Fiscal Years 2014–2015*, Washington, D.C., November 2016.

[11] Jeffrey M. Jones, "In U.S., Telecommuting for Work Climbs to 37%," *Gallup News*, August 19, 2015.

[12] Lister, 2016.

[13] Jones, 2015.

[14] Lister, 2016.

[15] U.S. Department of Transportation, *Transportation Implications of Telecommuting*, Washington, D.C., 1993.

exponentially workers' capabilities to work remotely.[16] Based on these factors, researchers since the 1980s have routinely overestimated the growth of teleworking. In 1991, telecommuting was projected to grow to between 12 and 25 million people by 2002.[17] By 1993, the estimate for teleworkers in 2002 was down to 7.5 to 15.0 million.[18] As early as 1995, researchers noticed that the "gap between expectations and reality has warranted deeper analysis as to why people refrain from telecommuting, even when seemingly offered the opportunity to do so."[19]

Costs and Benefits of Telework

The TEA requires annual assessments of agencies to show they have met their yearly outcome goals, but it does not require explicit reporting on cost savings. In 2016, GAO published a report on federal telework that showed that, even though OPM collects annual telework data across agencies, there were substantial declines in the reporting of telework cost savings from 2012 to 2013. GAO found that the guidance that OPM provides to agencies in support of their telework programs "lacks information about how agencies can use existing data collection efforts to more readily identify benefits [and costs] of their telework programs."[20] This further emphasized the lack of available information on the costs and benefits of telework. We learned that agencies are best able to produce quantifiable outcomes when they consider costs and establish measurable goals at the inception of their telework programs (see Chapter Three).

---

[16] Tuncay Bayrak, "IT Support Services for Telecommuting Workforce," *Telematics and Informatics*, Vol. 29, No. 3, 2012.

[17] Jack M. Nilles, *Telecommuting Forecasts*, Los Angeles, Calif.: Telecommuting Research Institute, 1991.

[18] U.S. Department of Transportation, 1993.

[19] Mokhtarian and Salomon, 1997.

[20] GAO, *Federal Telework: Better Guidance Could Help Agencies Calculate Benefits and Costs*, Washington, D.C., GAO-16-551, July 2016, p. 14.

The 2016 GAO report shows that the most commonly cited benefits of telework are "reduced employee absences, improved work/life balance, improved recruitment and retention, maintaining continuity of operations (COOP) during designated emergencies or inclement weather, reduced commuting costs/transit subsidies, increased productivity, reduced real estate costs, reduced utilities, and positive environmental impacts, such as reduced greenhouse emissions."[21] A number of studies summarize the hypothetical costs and benefits of teleworking, primarily based on the cost and benefit framework first proposed by Nilles and colleagues in 1976.[22] This framework hypothesizes cost and benefits at three levels: the employee, the employer, and the environment (see Table 2.1).

The following sections provide further detail on the costs and benefits reported in previous studies on telework within the framework shown in Table 2.1.[23] We do not attempt to be comprehensive, only to provide an overview of factors that may influence or affect employees, employers, and the environment.

---

[21] GAO, 2016, p. 5.

[22] Jack Nilles is credited with performing the first study on telecommuting and telework and is credited as the originator of the terms *telework* and *telecommute* (Nilles et al., 1976). The general taxonomy of costs and benefits at the employee, employer, and societal levels that he proposed in 1976 is still in use today, although his methods of projecting the impact of telework have been replaced with new models that better incorporate facilitators and barriers to telework. To facilitate telework, early models assumed extensive use of community "telework centers," which would bring high-speed computers and communications to teleworkers (Nilles et al., 1976; George S. Park, Jack M. Baer, and Walter S. Nilles, *Trends and Factors Influencing Telecommuting in Southern California*, Santa Monica, Calif.: RAND Corporation, DRU-1465-SCTP, 1996). The rapid growth of computers and communications infrastructure to facilitate telework means that most telework today is performed from home.

In terms of barriers to telework, there are a number of reasons workers choose to commute to a central office that early modelers did not consider, including the ability to separate their work from home life and to collaborate on a face-to-face basis with coworkers and supervisors. Management has also failed to embrace telework at the levels Nilles originally hypothesized—there is much disagreement in the literature as to why.

[23] A 2015 survey of the scientific literature on telework contains more than 200 citations for telework from the fields of psychology, management, transportation, communications, and technology, "the results of which are often conflicting" (Allen, Golden, and Shockley, 2015, p. 41).

Table 2.1
Hypothesized Benefits and Costs of Teleworking

Level	Benefit	Cost
Employee	• Direct savings in transportation costs and time • Indirect savings that derive from the ability to optimize where and when work activities are performed • Work-life balance	• Direct costs of maintaining a home office • Indirect costs associated with the professional and social impact of being isolated from the office • Indirect costs of needing to balance work and family
Employer	• Reduced employee absences • Direct savings from reduced real estate and other overhead costs • Indirect savings from higher worker productivity, better retention, the ability to recruit individuals independent of their places of residence • Indirect savings from maintaining COOP during designated emergencies or inclement weather	• Direct costs associated with fielding the infrastructure that makes telework possible • Potential direct costs of losing proprietary information, secrets, and data • Potential indirect costs associated with loss of line-of-sight supervisory control of employees • Potential costs in loss of the innovation and agility from ad hoc employee interactions
Environment	• Direct savings from wear and tear on roads • Indirect savings from better health from lower air pollution or the production of greenhouse gasses	• These costs were not considered in any of the models we reviewed

SOURCE: Nilles et al., 1976, supplemented by RAND analysis based on literature review.

Impacts on Employees
Benefits

The savings in commuting time—independent of how large or small that time is[24]—and the scheduling flexibility appear to be driving fac-

---

[24] P. Peters, K. Tijdens, and C. Wetzels, "Employees' Opportunities, Preferences, and Practices in Telecommuting Adoption," *Information and Management*, Vol. 41, No. 4, 2004.

tors for those who pursue telework.[25] The association between schedule flexibility and both the motivation to telework and the beneficial outcomes of telework is so strong that "the extent to which telecommuting is associated with beneficial outcomes may depend on the level of scheduling flexibility."[26] This schedule flexibility could be especially motivating for employees with additional family obligations. For instance, our review showed that being married and having children under 12 is one of the most consistent predictors of willingness or desire to telework among both genders.[27] Although the studies have not shown a statistically significant association between commute times and willingness to telecommute, the travel distance to work is often cited as a determining factor in an employee's decision to telework.[28]

Studies examining the costs and benefits of teleworking on the employees who participate in these programs find that those who telework are more engaged in their work and have greater job satisfaction than those who do not telework.[29] Social scientists posit that greater worker engagement might be the result of the greater autonomy felt by these employees or of a sense of gratitude for the flexibility to balance work and home life that teleworking provides.[30] However, it could be

---

[25] B. Baltes, T. Brigges, J. Huff, J. Wright, and G. Neuman, "Flexible and Compressed Workweek Schedules: A Meta-Analysis of Their Effects on Work Related Criteria," *Journal of Applied Psychology*, Vol. 84, No. 4, 1999.

[26] Allen, Golden, and Shockley, 2015.

[27] Diane E. Bailey and Nancy B. Kurland, "A Review of Telework Research: Findings, New Directions and Lessons for the Study of Modern Work," *Journal of Organizational Behavior*, Vol. 23, 2002; Y. D. Popuri and C. R. Bhat, "On Modeling Choice and Frequency of Home-Based Telecommuting," *Transportation Research Record*, Vol. 1858, 2003.

[28] Mokhtarian and Salomon, 1997; Park, Baer, and Nilles, 1996.

[29] Nilles et al., 1976; James G. Caillier, "Satisfaction with Work-Life Benefits and Organizational Commitment/Job Involvement: Is There a Connection?" *Review of Public Personnel Administration*, Vol. 33, No. 4, 2013b; Amanda J. Anderson, Seth A. Kaplan, and Ronald P. Vega, "The Impact of Telework on Emotional Experience: When, and for Whom, Does Telework Improve Daily Affective Well-Being?" *European Journal of Work and Organizational Psychology*, Vol. 24, No. 6, 2015.

[30] James G. Caillier, "The Impact of Teleworking on Work Motivation in a U.S. Federal Government Agency," *American Review of Public Administration*, Vol. 42, No. 4, 2012.

equally true that the most-engaged employees are simply the ones who telework; because supervisory permission is often needed to telework, it seems reasonable that supervisors might authorize telework only for their most engaged and trusted employees. In fact, Bailey and Kurland said that "little clear evidence exists that telework increases job satisfaction and productivity, as it is often asserted to do."[31]

Costs

The direct costs to employees of setting up home offices or financing telecommunications infrastructure—e.g., telephone and internet—do not appear to be significant factors in determining who will telework and how often. Prices for both computers and telecommunications infrastructure continue to fall while capacity increases, driven by consumer demand for internet content for everything from shopping to watching the latest movie releases. Additionally, more people already possess the required hardware and connectivity in their homes. The dual use of home computers and infrastructure for both personal business and work has instead produced a new security risk for the employer, a factor that early studies did not incorporate.

The indirect costs to employees who telework are assumed to be associated with isolation, both professional and social.[32] Professionally, there is a fear that *out of sight* means being *out of mind* when employers decide on promotions or who gets high-profile work assignments. Socially, employees may be adversely affected by the loss of informal interactions with coworkers. Despite existing studies showing that isolation is not a significant factor in the telework experience,[33] these fears continue to be reported as dominant factors in telework decisions.[34] Whether these fears actually motivate employee decisionmaking about whether to telework is unclear.

---

[31] Bailey and Kurland, 2002.

[32] Nancy B. Kurland and Terri D. Egan, "Telecommuting: Justice and Control in the Virtual Organization," *Organization Science*, Vol. 10, No, 4, 1999.

[33] Park, Baer, and Nilles, 1996; Kurland and Egan, 1999.

[34] Patricia Reaney, "About 20 Percent of Global Workers Telecommute: Poll," Reuters, January 24, 2012.

Telecommunications advances now allow managers, supervisors, and coworkers to monitor both the outputs and the online activities of workers. Workplace collaboration tools, such as Microsoft's Skype for Business and Cisco's Jabber, provide visibility of a colleague's online status and, when integrated with an employee's calendar, provide visibility to task lists and meeting schedules. These tools also provide voice, camera, and desktop-sharing services, allowing colleagues to discuss and review their work at any time, with just a click of a button. Teleconferencing tools that support impromptu meetings, such as Google Hangouts, are also growing in popularity. These advances in technology mean that teleworkers are no longer out of sight of colleagues. Although teleworkers might be less likely to suffer the negative side effects of isolation, advances in technology may also make it harder to reap benefits, such as the increased ability to concentrate or to escape office politics when working from home.

Another line of research examines the effect or indirect cost that telework has on allowing employees to balance work and family. Initially, research focused on the benefits of telework to avoid work interference with family, but studies that are more recent have examined how high-frequency telework can actually increase work interference with family. As early as 1999, some teleworkers reported "increased levels of overwork, invasion of personal life and loss of confidentiality" as significant costs.[35] There is also some evidence that for women who telework, domestic issues may negatively interfere with work and that telework may reinforce gendered division of labor within the home.[36]

Impacts on Employers
Benefits

The most substantial and direct benefit of telecommuting for employers occurs in reduced real estate and other overhead costs. However, these

---

[35] A. Pinsonneault, *The Impacts of Telecommuting on Organizations and Individuals: A Review of the Literature*, Montreal: HEC Montreal, 1999.

[36] Leslie Hammer, M. Neal, J. Newsom, K. Brockwood, and C. Colton, "A Longitudinal Study of the Effects of Dual-Earner Couples' Utilization of Family-Friendly Workplace Supports on Work and Family Outcomes," *Journal of Applied Psychology*, Vol. 90, No. 4, 2005.

savings can vary based on employers' use of arrangements that reduce the number of desks and parking spaces required. The most substantial indirect benefit to employers cited in the studies we reviewed is that of increased productivity for those who telework. A RAND report published in 1996 found that telework improvements in productivity were "due largely to a combination of reduced interruptions, higher ability to concentrate on the tasks at hand, more time spent actually working, and greater focus on performance."[37] Reports of higher productivity continue to dominate telework studies using both self-reported and supervisor-reported metrics.[38] However, many have pointed out the obvious: If only highly productive workers are allowed to telework, then one would expect those workers to be more productive. Additionally, if workers choose to perform tasks at home that require concentrated effort, which is a commonly cited motivator, both employers and employees may view those days as more productive than days spent at the office gathering the information to facilitate those efforts.[39]

Studies have also found that employees who telework stay with their employers longer than those who do not.[40] The 2016 GAO report attributed improved recruitment and retention to the implementation of telework at certain government agencies.[41] However, researchers noted that empirical research on this topic is limited and that most study conclusions are based on other factors, such as employee engage-

---

[37] Park, Baer, and Nilles, 1996.

[38] Stephen Ruth and Imran Chaudhry, "Telework: A Productivity Paradox?" *IEEE Internet Computing*, Vol. 12, No. 6, 2008; Ronald P. Vega, Amanda J. Anderson, and Seth A. Kaplan, "A Within-Person Examination of the Effects of Telework," *Journal of Business and Psychology*, Vol. 30, No. 2, 2015.

[39] Ralph D. Westfall, "Does Telecommuting Really Increase Productivity? Fifteen Rival Hypotheses," *AMCIS 1997 Proceedings*, Association for Information Systems Electronic Library, 1997.

[40] James G. Caillier, "Are Teleworkers Less Likely to Report Leave Intentions in the United States Federal Government Than Non-Teleworkers Are?" *American Review of Public Administration*, Vol. 43, No. 1, 2013a; Stewart D. Friedman and Alyssa Westring, "Empowering Individuals to Integrate Work and Life: Insights for Management Development," *Journal of Management Development*, Vol. 34, No. 3, 2015.

[41] GAO, 2016, p. 3.

ment and provision of greater flexibility, rather than actual retention rates.[42]

Job satisfaction does appear to be higher among teleworkers, with some researchers finding that this is related to the absence of interoffice politics and others finding that such satisfaction is associated with the greater flexibility that teleworking provides.[43]

Costs

Direct costs for employers to initiate a telework program include one-time costs for IT and other infrastructure setup, as well as ongoing costs for training and managing the telework program. However, as with the direct costs to employees, the direct costs for employers to set up such infrastructure were mitigated by the dramatic cost reductions in computer and telecommunications equipment over the past 20 years and do not appear to be a factor in private corporations' decision-making on telework.

Nevertheless, the direct costs of losing proprietary information, secrets, and data have become a significant concern for employers since Nilles first hypothesized the costs and benefits of telework.[44] If telework infrastructure provides additional entry points for cyber criminals or is implicated in a data breach, the result is a direct cost to the employer. Research indicates that possible loss of information is a growing concern for employers when making telework decisions, and the National Institute of Standards and Technology is updating its telework guidance to address cybersecurity concerns.[45]

---

[42] Allen, Golden, and Shockley, 2015.

[43] Kathryn L. Fonner and Michael E. Roloff, "Why Teleworkers Are More Satisfied with Their Jobs Than Are Office-Based Workers: When Less Contact Is Beneficial," *Journal of Applied Communication Research*, Vol. 38, No. 4, 2010; Anderson, Kaplan, and Vega, 2015; V. J. Morganson, D. A. Major, K. L. Oborn, J. M. Verive, and M. P. Heelan, "Comparing Telework Locations and Traditional Work Arrangements: Differences in Work-Life Balance Support, Job Satisfaction, and Inclusion," *Journal of Managerial Psychology*, Vol. 25, No. 6, 2010.

[44] Nilles et al., 1976.

[45] Mohana Ravindranath, "Is Telework a Growing Cyber Threat? New Guidelines Offer Security Tips," *Nextgov*, March 16, 2016.

The indirect costs to employers in terms of loss of knowledge sharing between individuals—especially loss of mentors, if senior employees telecommute frequently—and loss of innovation have been examined, but we were unable to find substantive research in these areas.[46] Substantive studies on the factors that promote innovation are especially lacking, although many researchers and firms believe that face-to-face contact is critical.[47] This poses real concerns to managers of creative work, but we found little empirical research on the topic despite Yahoo's high-profile decision in 2013 to rescind telework agreements to cultivate innovation.[48]

Impacts on the Environment
Benefits
In terms of benefits, the original research on telework assumed that reducing commutes to the office would result in a reduction in the annual national number of passenger vehicle miles.[49] However, empirical studies have shown only weak effects[50] or negative effects on vehicle miles.[51] The evidence suggests that telecommuters use their vehicles to make trips that previously may have been combined with the commute to work, offsetting the commute savings.

---

[46] L. Taskin and F. Bridoux, "Telework: A Challenge to Knowledge Transfer in Organizations," *International Journal of Human Resource Management*, Vol. 21, No. 13, 2010.

[47] J. Sullivan, "How Yahoo's Decision to Stop Telecommuting Will Increase Innovation," ERE Media, February 26, 2013.

[48] Dominic Basulto, "The Yahoo Memo and Marissa Mayer's Big Innovation Gamble," *Washington Post*, February 28, 2013.

[49] Nilles et al., 1976; U.S. Department of Transportation, 1993; Park, Baer, and Nilles, 1996.

[50] S. Choo, P. Mokhtarian, and I. Salomon, "Does Telecommuting Reduce Vehicle Miles Traveled? An Aggregate Time Series Analysis for the U.S.," *Transportation*, Vol. 32, No. 1, 2005.

[51] P. Zhu, "Are Telecommuting and Personal Travel Compliments or Substitutes?" *Annals of Regional Science*, Vol. 48, No. 2, 2012.

Costs

Environmental costs were not considered in any of the models reviewed, since we were unable to find data on environmental costs, such as greenhouse gas emissions avoided from teleworking.

Enabling Telework

The most frequently cited factors that enable the adoption and growth of telework are management or supervisor support and technology. Lack of manager support has been consistently reported as a significant barrier to telework adoption.[52] Managing teleworkers is thought to be a significantly different skill set from managing in-office employees;[53] performance must be judged on the output of the teleworkers rather than on their observed activities in the workplace. However, a few studies noted that all workers should be judged on both output and objective measures of activity and recommended that teleworkers be treated no differently from office workers.[54]

Conclusions

The literature related to telework is extensive, but we did not find consensus regarding what constitutes telework. This lack of consensus limited our ability to estimate trends in the growth and adaptation of telework across the United States. More widely adopted definitions, a clear differentiation between telework and telecommuting, and standard measures for time spent at an alternative work location versus

---

[52] Park, Baer, and Nilles, 1996; Mokhtarian and Salomon, 1997; Pinsonneault, 1999; Bailey and Kurland, 2002.

[53] S. Yu, "How to Make Teleworking Work," *Communications News*, Vol. 45, No. 12, 2008; Timothy R. Dahlstrom, "Telecommuting and Leadership Style," *Public Personnel Management*, Vol. 42, No. 3, 2013.

[54] Pinsonneault, 1999; B. A. Lautsch, E. E. Kossek, and S. C. Eaton, "Supervisory Approaches and Paradoxes in Managing Telecommuting Implementation," *Human Relations*, Vol. 62, No. 6, 2009.

time spent anywhere the employee desires to work would all provide a better understanding of the current and future national trends of telework across industry sectors.

Overall, we did not find consensus in the literature on the benefits and costs of telework, but our review identified benefits and costs to supplement the framework that Nilles laid out.[55] The literature provides examples of the types of benefits, measures, and cost savings that organizations may consider and plan for when designing a telework program. For employees, telework benefits may include greater flexibility in their schedules and reduced commuting time, in addition to the risk of isolation from the team environment that an office may provide. For employers, benefits may include increased productivity from employees and higher job satisfaction, although telework may cause decreased team cohesion. The most-significant monetary costs for employers may result from providing computing equipment or other infrastructure to employees working remotely, but there may also be some cost savings from decreasing the square footage of office space.

Chapter Three will further discuss costs and benefits as we examine the implementation of telework programs in government, specifically across seven federal agencies. These agencies share a consistent federal definition for *telework* and illustrate how the decision to implement telework in the federal government includes the consideration of additional factors not easily quantified or measured from one organization to the next.

---

[55] Nilles et al., 1976.

CHAPTER THREE

Examining Federal Telework Programs

The TEA, signed into law by President Barack Obama on December 9, 2010, was an important milestone in the history of telework across all federal executive agencies.[1] This law "specifies roles, responsibilities, and expectations for all Federal executive agencies with regard to telework policies; employee eligibility and participation; program implementation and reporting."[2] OPM responded by issuing a telework manual for executive agencies that suggested that all employees were eligible to telework unless their job titles required them to be on site every day or handle secure materials, they were otherwise not able to access their work remotely, or they had received a performance review below fully successful—or the agency equivalent—or any disciplinary action within the past 12 months.[3] Because federal agencies differ in their position classifications, organizational structures, mission areas, and other circumstances, "agencies have discretion to make their own eligibility determinations for employees subject to operational needs while considering the specific requirements described in the Act."[4] Therefore, different government agencies can specify their telework participation criteria and implementation policies.

As an attempt to understand how different agencies transformed their business practices to implement telework, we selected seven fed-

---

[1] Pub. L. 111-292, 2010.

[2] OPM, *Guide to Telework in the Federal Government*, Washington, D.C., April 2011.

[3] OPM, 2011.

[4] OPM, 2011, p. 14.

eral agencies that met the following criteria: a widely adopted telework program, publicly available data on the telework program, and responsibility for handling sensitive data. The seven agencies selected were

- Federal Emergency Management Agency (FEMA)
- General Services Administration (GSA)
- Internal Revenue Service (IRS)
- National Aeronautics and Space Administration (NASA)
- National Science Foundation (NSF)
- Nuclear Regulatory Commission (NRC)
- United States Patent and Trademark Office (USPTO).

We relied on publicly available data to understand why each agency implemented a telework program, how it implemented the program (which steps were taken), any investment costs or ROI it reported as a result of telework, and lessons learned from its experiences. We began our research with the OPM annual telework reports to identify agencies that report having active telework programs. We narrowed down the list of agencies to those that had both a significant number of teleworkers (relative to other agencies) and that handle sensitive information (such as nuclear information, classified information, or other protected types of data), and, finally, we researched which of those agencies publicly report data such that we could adequately conduct our analysis.

We sought agencies with active telework programs and that release publicly available data on the programs' benefits and costs in terms of mission impact, value to customers, finances, human capital metrics, and technology. We searched for data on performance measures that demonstrate the agencies' ROIs for changes made to implement their telework programs. These may include mission performance indicators, personnel retention or attrition, workforce morale, reduction in the real estate footprint of agency facilities, and financial savings. The degree to which agencies document qualitative and quantitative measures varies, and only some agencies release this information publicly. Therefore, we include exact figures and numbers only when they are available.

We looked for attributes of successful telework programs across four categories chosen by NGA: technological, legal, policy, and financial. For each category, we studied the attributes that make a telework program successful, and we show these attributes in Table 3.1. For technological attributes, we found that agencies that dedicate resources to their secure unclassified networks were able to increase the number of positions eligible to telework. Providing technical IT support and equipment also facilitated and ensured effective teleworking capabilities. Under legal attributes, we found that compliance with the TEA promoted the successful implementation of telework programs by providing a framework for implementation. Effective policy allowed for proficient performance management, adequate training for both employees and managers, and an overall high proportion of eligible teleworkers. Finally, in terms of financial attributes, agencies that successfully executed telework were able to reduce costs and have high ROIs.

In the following sections, we examine each agency and explain why the agency implemented telework, how it was implemented, the available data on investment costs and ROI, and lessons learned. Table 3.2 shows the most recently available data for telework at each of the seven agencies.

Federal Emergency Management Agency

FEMA's mission is to "support citizens and first responders to promote that as a nation we work together to build, sustain, and improve our capability to prepare for, protect against, respond to, recover from, and mitigate all hazards."[5] FEMA has ten regional offices throughout the United States.[6]

In fiscal year (FY) 2015, FEMA had 13,706 employees—10,280 were eligible to telework, and 6,247 teleworked. Therefore, approximately 46 percent of employees teleworked and 61 percent of eligible

---

[5] FEMA, "About the Agency," webpage, last updated May 16, 2017.

[6] Interview conducted by Maxwell students, May–June 2016.

Table 3.1
Attributes of Successful Telework Programs

Category	Attribute	Definition
Technological	Security	Agencies provide teleworkers with security measures to access their work remotely. These include virtual private networks (VPNs), encryption, password-protected websites, smart cards, and other safe means to access information.
	IT support	Agencies provide clear access to on-site and off-site technological support to thwart software or hardware telework malfunction. IT support is also available to provide employees with proper technology training (see next row).
	Equipment	Agencies provide the necessary equipment for employees to telework. The type of equipment ranges in scope and varies by agency but typically includes software and hardware.
Legal	Compliance with the TEA	Agencies comply, in whole or in part, with the TEA. Compliance includes, but is not limited to, the establishment of a clear telework policy, a written telework agreement between employees and managers, interactive telework training for employees and managers, the collection of telework data, and annual reporting to OPM.
Policy	Performance management	Agencies provide clear and defined methods for managers to track teleworkers' performances. These include, but are not limited to, virtual time management, productivity trackers, and virtual meetings to ensure that employees are completing their work.
	Training	Agencies routinely provide telework training to both managers and employees. This includes training that is mandated by agencies; successful agencies have implemented additional training, such as IT, communications, and management training.
	Proportion of teleworkers	Agencies have significant proportions of eligible employees who telework.
Financial	ROI	Agencies have significant ROIs.

Table 3.2
Status of Telework at Seven Agencies (FYs 2014–2015)

Agency	FY	Number of Employees			% of Eligible Employees Who Telework	% of All Employees Who Telework	Employees Who Telework . . .				
		Total	Eligible to Telework	Teleworkers			3 or More Days per Two-Week Period	1 to 2 Days per Two-Week Period	Once a Month	Situationally	Remote Workers
FEMA	2014	14,425	10,818	6,450	60	45	1,475 (10%)	2,482 (17%)	437 (3%)	2,064 (14%)	No data
	2015	13,706	10,280	6,247	61	46	2,170 (16%)	1,969 (14%)	377 (3%)	1,707 (12%)	24 (0%)
GSA	2014	11,506	10,597	9,711	92	84	5,067 (44%)	2,484 (22%)	345 (3%)	1,815 (16%)	No data
	2015	11,171	10,365	9,848	95	88	5,378 (48%)	2,433 (22%)	317 (3%)	1,720 (15%)	274 (2%)
IRS	2014	91,018	41,882	42,470	101	47	19,692 (22%)	4,529 (5%)	10,077 (11%)	—	No data
	2015	84,009	42,398	37,738	89	45	21,326 (25%)	4,152 (5%)	9,556 (11%)	—	36 (0%)
NASA	2014	18,493	17,951	9,088	51	49	28 (0%)	37 (0%)	28 (0%)	7,294 (39%)	No data
	2015	18,093	17,527	9,887	56	55	18 (0%)	34 (0%)	39 (0%)	7,996 (44%)	—
NSF	2014	1,432	1,271	1,151	91	80	293 (20%)	262 (18%)	478 (33%)	996 (70%)	No data
	2015	1,451	1,271	1,193	94	82	327 (23%)	250 (17%)	459 (32%)	1,046 (72%)	5 (0%)
NRC	2014	3,891	3,891	2,000	51	51	339 (9%)	449 (12%)	—	—	No data
	2015	3,810	3,810	2,200	58	58	385 (10%)	509 (13%)	—	—	68 (2%)
USPTO	2014	12,568	11,441	9,650	84	77	5,127 (41%)	4,305 (34%)	—	218 (2%)	No data
	2015	12,623	11,734	10,410	89	82	5,660 (45%)	4,454 (35%)	—	296 (2%)	2,043 (16%)

SOURCE: OPM, 2016.

NOTES: "No data" = data on remote workers were not asked for nor collected in FY 2014; — = no data reported.

employees teleworked.[7] There were 2,170 (16 percent of all employees) FEMA employees who teleworked three or more days per two-week period, 1,969 (14 percent) who teleworked one to two days per two-week period, 377 (3 percent) who teleworked once a month, and 1,707 (12 percent) who teleworked situationally.[8]

Why Was Telework Implemented?

FEMA cites several reasons for implementing telework, including that the telework program improves the life of the individual and the greater community by "reducing traffic congestion and pollution."[9] The agency's mission—to prepare for and respond to natural disasters and mobilize massive resources as necessary—is another incentive for the agency to be mobile. FEMA uses telework as a recruitment and retention tool to make positions within the component more attractive. It cites telework as directly driving the component's mission to maintain operations in the event of disaster, such as severe weather conditions.[10] The component's telework policies are implemented nationwide and utilized consistently in its regional offices.

According to FEMA, "Aligned with agency strategy and mission, telework supports achievement of objectives increasingly important for operation of an efficient and effective Federal Government, including cost savings and improved performance, and maximizing organizational productivity."[11] Tonya Schreiber, deputy chief administrative officer of the FEMA Mission Support Bureau, said, "We have a mobility mission. . . . Our stakeholders, federal partners and disaster survivors demand the ability for us to effectively deliver. FEMA goes

---

[7] OPM, 2016, p. 177.

[8] OPM, 2016, p. 201.

[9] FEMA, "Continuity Planning for Telework," Washington, D.C., undated-a.

[10] FEMA, undated-a.

[11] FEMA, undated-a.

big, and we go fast. We need to get in there, take charge, and leverage state and local partners, and we've got to be mobile across agencies."[12]

Employee buy-in is another important factor to enabling telework at FEMA. During Telework Week, a global event put together by the Mobile Work Exchange public-private partnership, FEMA had 66 percent workforce participation in 2013, the first year the agency participated. A report published by Mobile Work Exchange noted that as the agency becomes a mobile workforce, it is also "changing the tools to make employees more productive, efficient and responsive."[13] Nicole Early, telework managing officer and deputy chief component human capital officer at FEMA, reported, "[P]articipation in Telework Week showed great momentum in the strides we are making to meet mission requirements while helping employees to maintain a work-life balance."[14]

How Was Telework Implemented?

FEMA uses telework to ensure COOP during emergencies and provide flexible work options for employees. FEMA does not require teleworkers to use government-issued computers and software. However, if teleworkers do not currently own the required hardware to work from home, FEMA will provide the equipment and agency-standard software, such as Microsoft Office.[15] FEMA requires employees to have a Terminal Access Controller Access Control System (TACACS) account,[16] and the designated telework coordinator must approve access. This system allows teleworkers to remain online for the short time necessary to retrieve email.[17] If teleworkers in alternate work locations must access Sensitive But Unclassified (SBU) material and records that are subject

---

[12] Business of Federal Technology, "FEMA and Other Agencies Showcase Successful Mobility Initiatives," April 30, 2013.

[13] Mobile Work Exchange, *The Telework Revolution: Bringing Theory to Practice*, Alexandria, Va., 2013, p. 20.

[14] Mobile Work Exchange, 2013, p. 20.

[15] FEMA, *FEMA Manual 3000.3*, Washington, D.C., 2000, p. 16.

[16] FEMA, 2000, 15.

[17] FEMA, 2000.

to the Privacy Act of 1974, they must first be authorized by the office in charge of those records.[18]

Employees who wish to telework and their direct supervisors are required to attend training on policies, procedures, and organizational requirements specifically developed for the telework program.[19] FEMA also provides training for teleworking during emergencies—e.g., COOP trainings.[20] Aside from the required training, FEMA requires performance evaluations at designated milestones for employees who telework and consistent communication between managers and teleworkers to oversee performance.[21]

In the event of an emergency, FEMA has the COOP Telework Program in place, which allows employees to continue operations from alternate work sites. This plan has specialized training, and FEMA conducts COOP telework exercises to ensure that employees who support mission-essential functions are able to telework if a disaster occurs.[22] FEMA offers the following flexible telework options:

- regular telework pilot program—all official duties are performed at an alternate work site for up to three nonconsecutive days per week on a consistent basis
- episodic telework—all official duties are performed at an alternate work site for an agreed-upon period, typically for the duration of special projects
- medical telework—all official duties are performed at an alternate work site if an individual has disabilities that require such an accommodation or if an employee must care for an ill child or relative.[23]

---

[18] Public Law 93-579, Privacy Act of 1974, December 31, 1974; FEMA, 2000, pp. 18–19.

[19] FEMA, 2000, p. 9.

[20] FEMA, "Lesson 2: Elements of a Viable COOP," in *Continuity of Operations Planning Web-Based Course*, Washington, D.C., undated-b, p. 5.

[21] FEMA, 2000, p. 17.

[22] FEMA, *Department and Agency: Continuity Telework Exercise; Exercise Plan (EXPLAN) Template*, Washington, D.C., May 2013, p. 3.

[23] FEMA, 2000, p. 2.

FEMA's telework regulations allow most employees to be eligible for telework as long as their job responsibilities can be accomplished from an alternate location. FEMA managers recently responded to a telework survey, which found that approximately half of the managers in the sample regularly telework and that "70% have employees who currently telework at least one day per week; 56% of the managers responded that they were more productive when teleworking; and 30% noted higher performance in teleworking employees."[24]

Investment Costs and Return on Investment

Although specific data on the investment costs associated with implementing telework have not been published, multiple articles praise FEMA's high ROI. To achieve FEMA's goal of enabling employees to respond to disasters from anywhere, FEMA has closed five of its office buildings and "affirmed [its] commitment to a mobile workforce."[25] By increasing the number of employees who work from alternate locations, FEMA has saved $9.1 million a year in leasing costs and "roughly $530,000 per year in utility costs."[26] The agency was able to close these office buildings by creating open workspaces at the remaining offices, so that employees who regularly telework multiple days a week must share desks when they are working on-site.[27] FEMA, responding to a 2014–2015 OPM data call, stated that it decreased the amount of leased space by consolidating employees at its headquarters building. This was achieved, in part, by offering telework and employing desk-sharing.[28]

Lessons Learned

FEMA found that maintaining operations in the event of an emergency is essential to its agency mission. Integrating telework into the COOP program increased the number of employees who qualified for

---

[24] Jimmy Daly, "FEMA Is Ready for Anything—Even Telework," *FedTech*, May 24, 2013.

[25] Daly, 2013.

[26] Brittany Ballenstedt, "FEMA Ramps Up Telework, Mobility," *Nextgov*, May 10, 2013.

[27] Ballenstedt, 2013.

[28] OPM, 2016, p. 28.

telework. The agency also found that managers were changing telework agreements—particularly the days of the week when employees worked off-site—too frequently, sometimes on a weekly basis, which caused confusion for employees. Therefore, FEMA is currently working to revise the telework policy so that agreements remain consistent.[29]

General Services Administration

GSA supports the federal government by offering other government agencies products and services from commercial vendors, as well as the office space and facilities needed for agencies to better serve the public. GSA provides technological support to other governmental offices, preserves and maintains government buildings, and implements the American Recovery and Reinvestment Act of 2009 to transform federal buildings into high-performing green buildings.[30] GSA has 11 regional office buildings across the country, and its FY 2015 budget was $420 million for new projects and $965 million for reconstruction and maintenance efforts.[31]

In FY 2015, GSA had 11,171 employees: 10,365 were eligible to telework, and 9,848 teleworked. Therefore, approximately 88 percent employees teleworked and 95 percent of eligible employees teleworked.[32] There were 5,378 (48 percent) GSA employees who teleworked three or more days per two-week period, 2,433 (22 percent) who teleworked one to two days per two-week period, 317 (3 percent) who teleworked once a month, and 1,720 (15 percent) who teleworked situationally.[33]

---

[29] Interview conducted by Maxwell students, May–June 2016.

[30] Public Law 111-5, American Recovery and Reinvestment Act of 2009, February 17, 2009; GSA, "Background and History," webpage, last reviewed August 13, 2017a.

[31] GSA, "Regional Buildings Overview," webpage, last reviewed August 13, 2017b; Building Design and Construction, "House Passes 2015 GSA Budget with 17% Cut for New Construction Projects," July 23, 2014.

[32] OPM, 2016, p. 151.

[33] OPM, 2016, p. 161.

Why Was Telework Implemented?

GSA has been one of the federal agencies leading the transformation into a mobile workforce. The agency established its Mobility and Telework Policy in 2011 to comply with the objectives outlined in the TEA.[34] By increasing the number of teleworking employees, GSA sought to save money primarily on real estate, as well as on other standard workspace expenses, such as subsidizing employees' transit, stationery costs, and utility costs.[35] GSA also wanted to accommodate the changing demands of its workforce by providing employees the ability to work from home or an alternate, flexible physical and cultural space that allowed greater engagement and collaboration.[36] Therefore, telework was seen as a tool with the potential to benefit the employees and the employer.

How Was Telework Implemented?

Drafting the Mobility and Telework Policy was an evolving process. GSA employees are all eligible for telework, with the exception of the positions outlined in the TEA: positions that handle secure materials, including personally identifiable information (PII); positions that require on-site work that cannot be handled remotely; or employees who have been warned, reprimanded, or suspended for being absent without leave for more than five days or violating the Standards of Ethical Conduct for Employees of the Executive Branch in the past 12 months. Employees might also become ineligible for telework if doing so would result in diminished individual or organizational performance, or if they do not comply with the written telework agreement required between a supervisor and the employee. A key component of GSA's telework policy is that supervisors must maintain constant com-

---

[34] GSA, *GSA Telework Program Management Office: Recipe Book*, Washington, D.C., 2011, p. 5.

[35] GSA, "The Goals of Telework," webpage, last updated August 28, 2017f.

[36] GSA, 2017f.

munication and have frequent feedback meetings with their teleworking employees.[37]

Technology has been a particularly important factor in the efficiency and success of GSA's telework program. GSA's Program Management Office's Measurement Tranche and the Office of the Chief Information Officer partnered to develop interactive telework dashboards that allow agency-wide transparency of telework data by showing how many people telework and for how long.[38] The metrics embedded in these dashboards allow supervisors to track telework activity to identify employees who might be overworking or underworking or to compare how supervisors' programs and services are doing relative to those in other GSA regions. These dashboards also track information about how the agency is meeting goals and the agency's overall commuting costs and greenhouse gas emissions.[39] GSA provides computers to teleworking employees, VPN capabilities, and broadband cards to connect mobile devices to the internet regardless of location.[40] GSA identified the following technological capabilities as necessary for a successful telework program: (1) access to shared digital drives, databases, and regularly used computer programs from multiple locations; (2) an instant-messaging tool for quick, efficient online dialogue between colleagues; (3) remote email access hosted on an external digital server; (4) mobile phones that can be connected to a laptop; and (5) the creation and use of digital signatures in writing and PDF applications.[41]

All GSA employees who are eligible to telework may do so, regardless of position, as long as they pass an online training course that covers the agency's transformation to a more mobile workforce and "promotes enhanced customer relationships, stronger work/life bal-

---

[37] GSA, *General Services Administration (GSA) Information Technology (IT) Standards Profile*, GSA Order, CIO P 2160.1E, Washington, D.C., February 2014, p. 2.

[38] GSA, "Telework Dashboards," webpage, last reviewed August 13, 2017d.

[39] GSA, 2017d.

[40] GSA, *GSA Mobility and Telework Policy*, Washington, D.C., undated.

[41] GSA, "Resources for Managing Teleworkers," webpage, last reviewed August 13, 2017c.

ance, and the sustainability of a telework program."[42] Supervisors and managers are required to take additional training and may terminate teleworking privileges for individuals if there is a reduction in performance or productivity. However, the supervisor and the employee must address the problem and develop a plan that allows the employee to eventually return to teleworking. Ninety-six percent of GSA employees have completed the required training, and 95 percent of eligible GSA workers teleworked at some point during FY 2015.[43]

Investment Costs and Return on Investment

In FY 2010, 35 to 40 percent of GSA employees teleworked and helped save the agency $1.5 million.[44] By the end of 2013, GSA had shut down six office buildings and completely renovated its headquarters in Washington, D.C., to allow more-flexible workspaces.[45] Since the vast majority of GSA workers telework and, therefore, do not need an office, GSA was able to create an efficient and adaptive open floor plan: "The new mobile reconfiguration allowed an approximate increase of 60 percent of assigned workstations in the same area."[46] GSA's use of telework has also helped it reach its environmental goals. According to GSA, the agency reduced carbon dioxide emissions by 8,850 metric tons in 2013.[47] OPM commended GSA for its efforts to "reduce the footprint"[48]; part of those efforts resulted from a workplace transformation toward tele-

---

[42] GSA, *Telework Training*, Washington, D.C., last reviewed August 13, 2017e.

[43] Emily Kopp, "Don't Ask GSA How Many Full-Time Teleworkers It Has—It Doesn't Know," *Federal News Radio*, January 21, 2015; interview conducted by Maxwell students, May–June 2016.

[44] Matt McLaughlin, "Telework Thrives as Mobility Grows," *FedTech*, April 17, 2012.

[45] Casey Coleman, "Total Workplace Transforms Federal Office Space," *Around the Corner*, November 19, 2013.

[46] GSA, 2017e.

[47] OPM, *2014 Status of Telework in the Government Report to Congress: Fiscal Year 2013*, Washington, D.C., November 2015, p. 45.

[48] The GSA Carbon Footprint Tool was decommissioned in spring 2017 but was previously available at https://www.carbonfootprint.gsa.gov/.

work and a headquarters renovation that included hoteling.[49] These initiatives have increased building occupancy by 76 percent and reduced energy consumption by half.[50] This also accounted for savings of $24.6 million in rent and $6 million in administration costs.[51]

Teleworking also permits employees to continue to work during weather-related emergencies, which has saved GSA costs by allowing the agency to remain operational. For example, 4,000 GSA employees worked when many other federal agencies were forced to shut down during Hurricane Sandy in October 2012.[52] The agency was lauded for its telework program after Winter Storm Jonas in January 2016; the telework program allowed 3,600 out of 3,800 employees to continue to work through the storm.[53]

Lessons Learned

In 2015, GSA's telework program was audited by GSA's Office of Inspector General (OIG). Although the program saved the office money, the audit found that GSA was not accurately recording the number of teleworkers in the agency, the hours of full-time telework employees, or the number of employees who had completed the required training to become telework-eligible. Therefore, the OIG recommended that the agency revise some of its telework procedures and control.[54] GSA policy, which refers to full-time teleworkers as *virtual employees*, requires the Office of the Chief People Officer to track all virtual employees and virtual work arrangements. The GSA's Office of Human Resources provided us a list of 454 virtual and satellite employees (satellite employees work at GSA facilities, but not those of their own organizations); however, the OIG claimed the list was not accurate and that

---

[49] OPM, 2016, p. 22.

[50] OPM, 2016.

[51] OPM, 2016, p. 28.

[52] Joel Schectman, "GSA Enables More Telework," *Wall Street Journal*, September 20, 2013.

[53] David Shive, "Federal Workforce on Duty During Winter Storm Jonas," *Around the Corner*, General Services Administration, January 28, 2016.

[54] GSA, Office of Inspector General, *GSA's Program for Managing Virtual Employees and Teleworkers Needs Improvement*, Washington, D.C., 2015a, p. 5.

GSA could not identify which of those employees worked virtually and which were satellite employees.[55] Furthermore, the audit identified 19 employees who had recorded significant telework hours but were not included on the Office of the Chief People Officer's list.[56] The audit also suggested that travel costs related to virtual work arrangements and office duty stations were inaccurate, because they were not being assessed annually. In a statement, GSA's chief human capital officer concurred with the findings of the report, attributed the mistakes to inaccurate manager documentation, and stated that GSA would make revisions to its telework policy and correct program errors.[57]

According to an official at GSA, the agency has learned that employees and managers need to communicate more and that managers need to be more aware of worker productivity.[58] GSA learned that not everyone will embrace telework; however, it is still a great tool, and improvements to the program, such as through training and support to employees, should always be considered.[59]

Finally, GSA's efforts do not provide benefit solely to the agency. The U.S. Department of Education has noted increased telework, in part because of working with GSA in office-consolidation efforts, which allows more employees and managers to telework.[60]

Internal Revenue Service

The IRS operates under the U.S. Department of the Treasury and is responsible for the collection of individual income and employment taxes. The agency has four primary divisions: Small Business/Self-Employed, Wage and Investment, Tax-Exempt and Government Enti-

---

[55] GSA, Office of Inspector General, 2015a.

[56] GSA, Office of Inspector General, 2015a, p. 2.

[57] Kopp, 2015.

[58] Interview conducted by Maxwell students, May–June 2016.

[59] Interview conducted by Maxwell students, May–June 2016.

[60] OPM, 2016, p. 217.

ties, and Large Business and International; their respective headquarters are in Lanham, Maryland; Atlanta, Georgia; Cincinnati, Ohio; and Washington, D.C.[61]

In FY 2015, the IRS had 84,009 employees: 42,398 were eligible to telework, and 37,738 teleworked. Therefore, approximately 45 percent of employees teleworked, and 89 percent of eligible employees teleworked.[62] There were 21,326 (25 percent) IRS employees who teleworked three or more days per two-week period, 4,152 (5 percent) who teleworked one to two days per two-week period, and 9,556 (11 percent) who teleworked once a month.[63]

Why Was Telework Implemented?

Telework was first implemented around 1990 at the IRS to cut costs and improve productivity.[64] Saving money is part of the agency's mission, and, in FY 2015, the IRS spent 35 cents for every $100 collected in taxes.[65] The greatest cost savings the IRS has reaped from telework is in reduced rent payments, but other claimed or documented benefits to the agency include the "ability to continue agency work" in the case of an emergency, "enhanced recruitment and retention," greater "environmental sustainability," "increased employee productivity and job satisfaction," and "improved work-life balance."[66]

How Was Telework Implemented?

To be compliant with the TEA, the IRS *Internal Revenue Manual* requires the agency to notify employees of telework eligibility, to train

---

[61] IRS, "At a Glance: IRS Divisions and Principal Offices," webpage, last reviewed July 27, 2017a.

[62] OPM, 2016, p. 189.

[63] OPM, 2016, p. 213.

[64] Interview conducted by Maxwell students, May–June 2016.

[65] IRS, "The Agency, Its Mission and Statutory Authority," webpage, last reviewed August 6, 2017b.

[66] IRS, "Part 6: Human Resources Management, Chapter 800. 0 Employee Benefits, Section 2: IRS Telework Program," in *Internal Revenue Manuals*, Washington, D.C., last reviewed September 10, 2017c.

employees and managers about telework, to have each employee sign a telework agreement before participation, and to provide managers with the authority to terminate the telework agreement if productivity becomes insufficient.[67] The IRS telework program offers four types of telework arrangements: ad hoc, recurring, frequent, and full time.[68] Full-time telework is conducted from a location that is not an official IRS work site. Ad hoc and frequent telework do not occur on a recurring basis; ad hoc is short term or episodic, and frequent is more than 80 hours each month. Temporary telework is approved in cases of employee hardship and cannot exceed 120 days. It generally occurs when medical situations or family hardships arise. Telework is not permitted without a telework agreement in place.

Employees in a range of occupations are eligible for telework, including revenue agents, computer audit specialists, budget technicians, IT specialists, engineers and appraisers, attorneys, tax specialists, translators, tax analysts, budget analysts, statisticians, and procurement technicians.[69] Additional requirements for eligibility are at least one year of employment at the service, a "fully successful performance appraisal," no disciplinary or adverse actions that would affect integrity in the telework program, a telephone where the employee can be reached, and an adequate workspace with equipment suitable for the security and protection of government property.[70]

The IRS requires training for employees and managers.[71] Additionally, there are supplemental courses available about telework for those who want to further their knowledge; these courses are available through the Enterprise Learning Management System.[72] Telework at the IRS is documented through a Single Entry Time Reporting

---

[67] Telework.gov, "Telework Enhancement Act," webpage, undated.

[68] IRS, 2017c.

[69] IRS and National Treasury Employee Union, *2012 National Agreement II*, Washington, D.C., 2012, p. 147.

[70] IRS, 2017c.

[71] IRS, 2017c.

[72] IRS, 2017c.

(SETR) System.[73] Teleworkers are required to submit their time and attendance accurately to indicate their hours of telework in the SETR System.[74] Managers are required to verify employee time and attendance to confirm that the SETR System is used properly.

Pursuant to the *Internal Revenue Manual*, access to sensitive information while teleworking is required to be compliant with the Privacy Act.[75] To avoid compromising the requirements under the Privacy Act, employees are required to use passwords and to encrypt data. Certain information on hard copies must be locked in a filing cabinet when not in use; the IRS provides locked filing cabinets to teleworkers. IRS telework employees must also follow specific procedures when disposing of sensitive materials.

Investment Costs and Return on Investment

During FY 2013, the IRS closed 22 small offices, saving $410,539 in rent and utilities; these savings are visible in the leasing budget line of the FY 2013 budget.[76] The IRS keeps track of cost savings for office closures through a program named Home as Post of Duty. Telework evidently plays a large role in office closures. In a 2012 press release, the Treasury Inspector General for Tax Administration (TIGTA) projected that the IRS could potentially save more than $111 million in leasing costs between 2012 and 2017, because of increased workplace-sharing from telework, but no data have been released to confirm actual savings.[77] However, the Department of the Treasury, in response to a 2014–2015 OPM data call, stated that its three bureaus (the IRS is one) generated more than $5 million in savings, partially because of telework.[78]

---

[73] IRS, 2017c.

[74] IRS, 2017c.

[75] IRS, 2017c.

[76] OPM, 2015, p. 337.

[77] TIGTA, "TIGTA: IRS Can Reduce Real-Estate Costs by Increasing Use of Telework," press release, September 24, 2012b.

[78] OPM, 2016, p. 28.

The IRS invests in certain materials for employees who telework, including a lockable filing cabinet; calculator; computer; telephone; second telephone line; and the capability to print, scan, and copy as needed.[79] Actual figures for investment in the telework program were not available.

Lessons Learned

In 2012, the TIGTA recommended a workspace-sharing initiative to cut costs.[80] The TIGTA concluded that the IRS was saving money with the telework program but that further savings were possible. Many employees were teleworking between one and five days per week, leaving empty offices that could be used as shared workspace. The TIGTA argued that a workplace-sharing initiative would save additional leasing and utility costs. IRS management agreed with the TIGTA and signed into agreement a workspace-sharing program. After a successful negotiation with the National Treasury Employees Union, the IRS implemented the program on October 1, 2012.[81] The 2013 TIGTA audit of telework implementation further recommended that the IRS improve its remote-management system, arguing that telework employees should have the same expectations of work production, rules of conduct, time and attendance, ethics, and all other regulations as those not teleworking and applicable to their positions.[82]

In 2016, a TIGTA audit revealed that 35 employees (21 percent of a random sample of 165 employees) had teleworked despite not having a valid telework agreement on file. It also found that 103 of the approximately 37,000 employees who telework had been disciplined for serious misconduct, but only 6 percent of these employees had stopped tele-

---

[79] IRS and National Treasury Employee Union, 2012, p. 150.

[80] TIGTA, *Significant Additional Real Estate Cost Savings Can Be Achieved by Implementing a Telework Workstation Sharing Strategy*, Washington, D.C.: U.S. Department of the Treasury, August 27, 2012a, p. 8.

[81] TIGTA, 2012b.

[82] TIGTA, Office of Inspections and Evaluations, *Review of the Implementation of the Telework Enhancement Act of 2010*, Washington, D.C.: U.S. Department of the Treasury, July 2013, p. 13.

working after the misconduct was revealed. In that report, the TIGTA recommended "that the IRS Human Capital Officer: 1) finalize processes to reasonably ensure that teleworking employees have completed telework training and have a valid telework agreement on file prior to beginning telework, 2) clarify the rules associated with misconduct, and 3) develop procedures for identifying employees whose conduct may impact the integrity of the Telework Program."[83]

National Aeronautics and Space Administration

NASA is responsible for science and technology related to air and space. It has nine field centers; a headquarters in Washington, D.C.; and the shared services center in Mississippi.[84]

In FY 2015, NASA had 18,093 employees: 17,527 were eligible to telework, and 9,887 teleworked. Therefore, approximately 55 percent of NASA employees teleworked, and 56 percent of eligible employees teleworked.[85] There were 18 (0.1 percent) NASA employees who teleworked three or more days per two-week period, 34 (0.2 percent) who teleworked one to two days per two-week period, 39 (0.2 percent) who teleworked once a month, and 7,996 (44 percent) who teleworked situationally.[86]

Why Was Telework Implemented?

NASA implemented telework to improve employee work-life balance and to "enhance the recruitment and retention of a high-quality diverse workforce; assist in providing reasonable accommodations to individuals with disabilities, including employees who have temporary or continuing health conditions; provide for the continuity of opera-

---

[83] TIGTA, *Telework Qualification Requirements Are Generally Being Met, but Program Improvements Are Needed*, Washington, D.C.: U.S. Department of the Treasury, July 19, 2016.

[84] NASA, *FY 2015 Agency Financial Report*, Washington, D.C., 2015, p. 11.

[85] OPM, 2016, p. 152.

[86] OPM, 2016, p. 162.

tions during national or regional emergencies; reduce transportation-related stress and costs; improve morale by allowing employees to balance work and family demands; and encourage the highest employee productivity toward the accomplishment of the Agency's mission."[87] In implementing telework, NASA found that telework also improves employee health, creates staffing flexibility, enhances employee retention, preserves the environment, creates cost savings, and reduces the overall organizational carbon footprint.[88] NASA ranked highest among large federal agencies in 2014 for its employee satisfaction, attributed in part to the work-life balance offered with the teleworking opportunities, childcare services, and wellness programs.[89]

How Was Telework Implemented?

Employees who have job duties that may be completed at an alternate work site are eligible for telework agreements. However, employees with job duties that require hands-on activity at NASA office locations, those who handle classified materials on a regular basis, or those who received a "less than fully successful" performance rating or their conduct resulted in disciplinary action within the past 12 months are ineligible for telework agreements.[90] In addition, employees can telework during emergencies to ensure COOP.

NASA provides employees access to SBU information and PII during telework. NASA uses multiple software programs for teleworking, including Skype, Adobe Connect, Google Docs (the agency recently joined a pilot effort with Google Apps for capabilities specific to NASA needs), DropBox, and Asana ("a web task-management application that supports efficiency among distributed teams").[91] NASA

---

[87] NASA, *NASA Procedural Requirements*, Washington, D.C.: Office of Human Capital Management, NPR 3600.2, 2010b, p. 3.

[88] NASA, "Work from Anywhere: Factsheet," undated-b.

[89] Hannah Moss, "What Your Agency Can Learn from NASA," *Gov Loop*, December 11, 2014.

[90] NASA, *NASA Desk Guide on Telework Programs*, Washington, D.C.: Office of Human Capital Management, NSREF-3000-0012, April 2010a, p. 8.

[91] Ali Llewellyn, "Teleworking," *Open NASA*, November 15, 2011.

requires that employee telework arrangements be evaluated during annual performance reviews and that employees "have at least a performance summary rating of 'fully successful'" to continue to participate in the telework program.[92] NASA also requires that telework performance "be monitored in the same manner as that of an employee who is onsite."[93] These tools, in conjunction with required employee and supervisor training, allow for effective performance management within the organization.[94]

NASA has implemented methods, in addition to software, that allow employees to work on SBU and PII remotely while still maintaining the necessary levels of security. There is no conclusive evidence regarding the efficacy of NASA's approach to telework security—i.e., providing government-issued laptops and VPN capabilities.[95] The agency is also "moving towards more elaborate security approaches, such as card readers in addition to password protection."[96] For employees to access the NASA Headquarters Network through the VPN, they must have a government-issued computer with VPN software installed, a smartcard (NASA Badge), a PIN for logging into the computer and VPN access, a SecurID token, and an eight-digit alphanumeric PIN for access to token-only applications and websites.[97] In addition to these requirements, Apple computers must also have a SecurID token and eight-digit alphanumeric PIN for access to the VPN and an Agency User Identification username and NASA Data Center password.[98]

NASA offers three flexible telework options: full-time telework, where all official duties are performed at an approved alternate work

---

[92] NASA, 2010a, p. 10.

[93] NASA, 2010a.

[94] NASA, *Work from Anywhere: NASA's Telework Program*, Stennis Space Center, Miss.: NASA Shared Services Center, March 2013.

[95] Georgetown University Law Center, *Telework in the Federal Government: The Overview Memo*, Washington, D.C., 2009.

[96] Georgetown University Law Center, 2009.

[97] NASA, "Remote Access Services," webpage, Information Technology and Communication Division, undated-a.

[98] NASA, undated-a.

site; part-time telework, where official duties are performed at an approved alternate work site for certain days of the week and in the office for other days of the week; and temporary telework, where official duties are performed at an alternate work site on an as-needed basis, such as emergency situations.[99]

Investment Costs and Return on Investment

Although certain conditions need to be met to ensure that the employee is successfully able to work away from the traditional work site before permission is granted, most positions at NASA are eligible for the Work from Anywhere program.[100] The Work from Anywhere program saved the federal government $30 million in productivity for each day the government closed because of the 2010 snowstorms.[101] The program enhances recruitment and retention of employees by providing them with "the same tools and online capabilities as best-in-class private sector organizations," allows increased productivity for the agency, minimizes costs, and helps reduce carbon emissions by limiting travel and reducing the need for new construction.[102] This program benefits employees, the agency, and society.

Lessons Learned

A key challenge to implementing telework for most agencies is the reluctance of management to allow employees to telework. This reluctance stems primarily from management anxiety about being unable physically to see employees working and, therefore, believing that they are unable to measure productivity and output. In the case of NASA, requiring teleworkers to use productivity tracking software, such as the aforementioned Asana software, reduced management anxiety about

---

[99] NASA, 2010a, p. 7.

[100] NASA, undated-b.

[101] NASA, undated-b.

[102] NASA, undated-b.

signing telework agreements and helped employees feel that their work was being acknowledged.[103]

National Science Foundation

NSF was created in 1950 as an independent federal agency that supports fundamental research and education and aims to "promote the progress of science; to advance the national health, prosperity, and welfare; and to secure the national defense."[104]

In FY 2015, NSF had 1,451 employees: 1,271 were eligible to telework, and 1,193 teleworked. Therefore, approximately 82 percent of NSF employees teleworked, and 94 percent of eligible employees teleworked.[105] There were 327 (23 percent) NSF employees who teleworked three or more days per two-week period, 250 (17 percent) who teleworked one to two days per two-week period, 459 (32 percent) who teleworked once a month, and 1,046 (72 percent) who teleworked situationally.[106]

Why Was Telework Implemented?

In 2008, NSF collaborated with Telework Exchange, a public-private partnership focused on telework, to study the foundation's telework program. The study surveyed employees, managers, and directors and, based on an 87 percent response rate, found telework to be a "win-win-win for employees, managers, and the environment."[107] Overall, the report found that, as of 2008, 51 percent of NSF employees participated in telework and 32 percent on a daily basis.[108]

---

[103] Llewellyn, 2011.

[104] NSF, "About the National Science Foundation," webpage, undated.

[105] OPM, 2016, p. 153.

[106] OPM, 2016, p. 163.

[107] NSF, *Telework Under the Microscope: A Report on the National Science Foundation's Telework Program*, Alexandria, Va., 2008b.

[108] NSF, 2008b.

NSF has sought to increase work-life balance by clarifying and enhancing its telework program. In 2011, NSF set a goal to create a more robust telework program as a tool to retain talented employees.[109] In 2014, NSF updated its human resources and telework policies to make telework more accessible and manageable for both managers and employees, including increasing the maximum number of days per week that an employee could telework. This new policy increased the total number of hours teleworked by 26 percent the first year after it was issued and attributed to some of the reduction in NSF's FY 2014 energy and water consumption.[110] NSF's 2015 sustainability report cited telework as a key driver of sustainable business practices. According to the report, telework contributed to a 13 percent drop in electricity consumption in FY 2013 and reduced the use of heat-generating computers and other equipment.[111] Telework is thus seen as a necessary tool to reach the agency's sustainability goals. These goals include reduced greenhouse gas emissions, sustainable buildings, and water-use efficiency and management.[112] The report also stated that telework participation is expected to continue rising as more employees and their supervisors become more comfortable with telework.

How Was Telework Implemented?

While there is little public information about NSF's initial transition to telework, clear guidelines and policies outline how telework is currently implemented across the agency. Since 2012, NSF has used an electronic telework application system (webTA) to track telework agreements throughout the foundation.[113] Telework agreements, as mandated in the TEA, are used to track and document telework hours. WebTA has been credited with giving the telework program "improved customer

---

[109] OPM, 2015, p. 83.

[110] NSF, *FY 2015 Strategic Sustainability Performance Plan*, Alexandria, Va., 2015, p. 1.

[111] NSF, 2015, p. 2.

[112] NSF, 2015, p. 3.

[113] OPM, 2015, p. 81.

service" and "program efficiency," and "the system facilitates tabulation of data to be used to identify trends and make data-driven decisions."[114]

To attract top talent, telework is consistently referenced as a workplace flexibility option for employees in NSF's job vacancy announcements. The chief of the Employee Relations Branch informs new employees of telework options on the second day of employment, during the new-employee orientation.[115] The agency has also held a series of comprehensive town halls on the implementation of its new telework policy program. These are used as a means to brief employees on policy system changes, answer questions, and achieve workforce buy-in.[116]

NSF is also working on initiatives to improve its telework program. It is currently developing change management plans, improving workforce communications, and updating required documents for system changes, as outlined in its new telework policy. As of 2013, NSF was developing supervisor telework training to allow managers to include telework as a metric of performance management. In 2014, NSF aligned its telework policy with the TEA. In addition to this and other policy issuances, NSF has improved communication to employees and the marketing of its telework policy. NSF has taken a collaborative approach, updated its website, and received and incorporated employee feedback.[117] Additionally, NSF's "2014 telework awareness week" offered training sessions to on-site staff and virtually accessible teleworkers.[118]

Investment Costs and Return on Investment
The *FY 2015 Strategic Sustainability Performance Plan* noted that telework reduced energy intensity, the most important factor reflecting an agency's sustainability, by 26 percent compared with 2008 levels. This

---

[114] OPM, 2015, p. 75.

[115] OPM, 2015.

[116] OPM, 2015.

[117] OPM, 2016, p. 245.

[118] OPM, 2016, p. 245.

reduction undoubtedly led to cost savings, although exact numbers are unavailable. The report also noted that water consumption decreased by more than 14 percent between FY 2013 and FY 2014 alone. Telework was highlighted as a key driver behind cost reduction and sustainability practices.[119]

A 2008 NSF report noted that, collectively, NSF employees who telework save more than $700,000 annually on commuting costs.[120] The report also indicated that 67 percent of employees believed that telecommuting increased their productivity, suggesting that teleworking has produced cost savings for both employers and employees.[121]

Lessons Learned

The success of NSF's telework initiative is largely the result of management support. As identified earlier, management support is a critical component of successful implementation of telework programs. In 2014, NSF's chief human capital officer, along with the director of human resource management, sent agency-wide guidance supporting and encouraging the new telework policy.[122] This allowed mobile work flexibility to those who asked for it and encouraged others to try it as well.

NSF has a telework managing officer who briefs customer organizations and facilitates discussions about how to work through concerns regarding telework. The officer encourages a collaborative approach between supervisors and employees to support increased telework usage and generally a more robust NSF telework program. NSF launched in July 2015 a "reconfigured electronic telework agreement tracking system" that complied with the foundation's new telework policy, featuring additional enhancements and flexibilities included in the policy. The NSF chief human capital officer was supportive of the efforts to

---

[119] NSF, 2015.

[120] NSF, "'Telework' Benefits Employers, Employees and the Environment," news release, March 11, 2008a; NSF, 2008b

[121] Rami Mazid, "Cisco Study Finds Telecommuting Significantly Increases Employee Productivity, Work-Life Flexibility and Job Satisfaction," Cisco, June 26, 2009.

[122] OPM, 2016, p. 245.

complete the system.[123] This shows the importance of a collaborative approach to ensure the success of telework in an organization.

Nuclear Regulatory Commission

The NRC was created by Congress in 1974 as an independent agency to regulate the commercial nuclear industry. It is responsible for regulating commercial nuclear power plants and other uses of nuclear materials, such as nuclear medicine, through licensing, inspection, and enforcement of the requirements to protect public health and safety against radiation hazards from industries that use radioactive material.[124] The NRC employs approximately 4,000 people.[125] The NRC has four regional offices and 65 nuclear sites, and the headquarters is in Rockville, Maryland. Figure 3.1 depicts full-time teleworker locations.[126]

In FY 2015, the NRC had 3,810 employees: 3,810 were eligible to telework, and 2,200 teleworked. Therefore, approximately 58 percent of NRC employees teleworked, and 58 percent of eligible employees teleworked.[127] There were 385 (10 percent) NRC employees who teleworked three or more days per two-week period and 509 (13 percent) who teleworked one to two days per two-week period; additional data for remaining teleworkers are not available.[128]

Why Was Telework Implemented?

According to a case study conducted by Commuter Connections, "NRC's telework program began in 1997 to allow employees to work

---

[123] OPM, 2016, p. 291.

[124] Computer Connections/Metropolitan Washington Council of Governments, *Employer Telework Case Study*, Washington, D.C., 2010; NRC, "About NRC," webpage, undated.

[125] Interview conducted by Maxwell students, May–June 2016.

[126] NRC, "Locations," webpage, last reviewed September 25, 2017.

[127] OPM, 2016, p. 153.

[128] OPM, 2016, p. 163.

Figure 3.1
NRC Full-Time Teleworker Locations

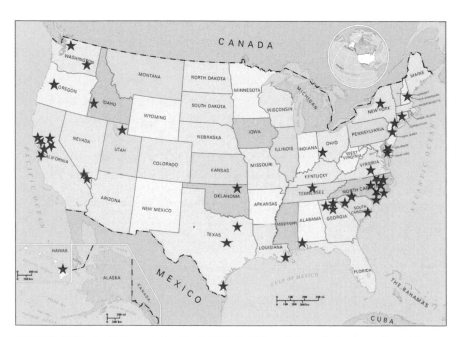

SOURCE: NRC, *Audit Report: Audit of NRC's Full-Time Telework Program*, Rockville, Md., OIG-14-A-05, December 2013.
RAND *RR2023-3.1*

away from the traditional office and provide a practical solution to environmental and other quality of life issues. Allowing employees to telework also helped address work life challenges."[129] After implementing its telework program, the NRC saw many benefits, including reduction in employee stress levels, reduced commuting time and costs, and freedom from office distractions.[130]

How Was Telework Implemented?

The NRC, according to one official, has a collective bargaining agreement for the entire organization that is negotiated between manage-

---

[129] Computer Connections/Metropolitan Washington Council of Governments, 2010, p. 1.

[130] Computer Connections/Metropolitan Washington Council of Governments, 2010.

ment and the union of representatives.[131] The NRC requires that a supervisor approve all telework in advance. In compliance with the TEA, incoming employees learn during orientation about the NRC's telework program. The telework program provides three options for employees: fixed schedule, project-based, and special circumstances. Moreover, the office of human resources meets annually with regional telework coordinators and briefs them on policies and procedures as they relate to telework.[132] NRC teleworkers are provided with Citrix Online remote access software to access the NRC intranet.

Employees can telework on a regular schedule under the fixed-schedule option. We were unable to find information about whether the telework centers are other government agencies' facilities with which the NRC has a sharing agreement. The project-based telework option allows employees to telework on an as-needed project basis, granted once employees have received approval from their supervisors. This option is usually allowed for projects that are required to be completed in a short period, typically from a few hours to a few days, that can be done at home or from a specific telework center.[133] Approximately 1,200 employees telework under this arrangement.[134] The NRC's telework program specifies that permission to telework may be granted "for special circumstances"—the third telework option. Employees who are incapacitated or are facing a significant hardship may be approved to telework continuously or intermittently.[135]

Teleworkers who work with sensitive material are required to get permission or a waiver prior to removing sensitive material from NRC facilities. If employees cannot get permission, they can work only on unclassified and nonsensitive material while out of the office. Employ-

---

[131] Interview conducted by Maxwell students, May–June 2016.

[132] Computer Connections/Metropolitan Washington Council of Governments, 2010, p. 2.

[133] Computer Connections/Metropolitan Washington Council of Governments, 2010.

[134] Interview conducted by Maxwell students, May–June 2016.

[135] Computer Connections/Metropolitan Washington Council of Governments, 2010, pp. 2–3.

ees who have access to sensitive information can work through the NRC's secure intranet, called Citrix Online.[136]

Investment Costs and Return on Investment

The NRC has not released information on investment costs or ROI, as its system is not currently automated. This makes it difficult for the NRC to make determinations without tracking the specifics of individual teleworkers.[137] Unlike other agencies, the NRC does not provide any equipment, such as computers, internet access, or mobile phone accommodations, to employees. We were unable to find public data on cost savings potentially generated from these and other areas, such as real estate.

Lessons Learned

A report conducted by the NRC's OIG in 2013 demonstrated that NRC's telework program showed opportunity for improvement with its training of employees and supervisors. For example, in 2013, 73 percent of full-time teleworkers and 80 percent of managers were not receiving adequate telework training. According to the TEA and the NRC's telework policy, a full-time telework employee is one who teleworks at least 90 percent of the time.[138] The NRC was not compliant with the TEA, because it did not require telework employees to complete the necessary training and did not have adequate tracking tools to assess employee productivity. According to the OIG audit, it is necessary to use a tracking tool to ensure that employees and supervisors have completed the necessary training and are productive while teleworking. Since the audit, the NRC has used the iLearn Learning Management System to ensure that all eligible telework employees and supervisors have completed all required training. IT has also been an issue when implementing the telework program, because it has been a challenge to provide employees who work from home with access to

---

[136] Interview conducted by Maxwell students, May–June 2016.

[137] Interview conducted by Maxwell students, May–June 2016.

[138] NRC, 2013, p. 3.

their records and with the information they need to complete their jobs.[139]

Furthermore, one official shared that the NRC has previously faced management buy-in and IT issues at various levels throughout the organization but that additional education and training, such as telework briefings and conferences, have allowed management to become more comfortable with the program.[140] Communication between the supervisor and employee is particularly important as the telework program becomes more popular, and employees need to be reminded what their roles and responsibilities are while teleworking. Through education and training, supervisors and employees can ensure that the necessary authorization is granted before teleworking begins and that expectations are understood. Subsequent to the audit, the NRC made strides to ensure that all eligible employees and supervisors were trained and supported when implementing telework. The OIG has since declared the NRC in full compliance with the TEA.[141]

United States Patent and Trademark Office

USPTO is responsible for granting U.S. patents and registering trademarks. The office promotes effective intellectual property protection for U.S. citizens who are innovators and entrepreneurs worldwide and provides training, education, and capacity programs designed to foster and encourage the development of intellectual property–enforcement regimes by U.S. trading partners.[142] USPTO was selected for this report as an example of an agency that implemented telework and encountered challenges with its program. We discuss some of the challenges in the section about lessons learned.

---

[139] Interview conducted by Maxwell students, May–June 2016.

[140] Interview conducted by Maxwell students, May–June 2016.

[141] GSA, Office of Inspector General, *Status of Recommendations: Audit of NRC's Full-Time Telework Program*, Washington, D.C., 2015b.

[142] USPTO, "About Us," webpage, undated.

In FY 2015, the USPTO had 12,623 employees: 11,734 were eligible to telework, and 10,410 teleworked. Therefore, approximately 82 percent of USPTO employees teleworked, and 89 percent of eligible employees teleworked.[143] There were 5,660 (45 percent) USPTO employees who teleworked three or more days per two-week period, 4,454 (35 percent) who teleworked one to two days per two-week period, and 296 (2 percent) who teleworked situationally.[144]

Lessons Learned

USPTO led a telework program that had been highly praised until fraud complaints by four whistleblowers to the OIG in the Department of Commerce prompted an investigation of the program in 2012. USPTO responded with a filtered response that omitted the complete findings of that investigation and reported instead that "interviews with the managers showed 'inconsistent' views on whether examiners were gaming the system."[145] However, the full investigation discovered that a portion of patent examiners had misrepresented the hours they were submitting and received bonuses for work they never did.[146] A second investigation that began shortly after found that dozens of paralegals, also working from home, had been paid full salaries during a four-year period where they had "limited assignments as they waited for new judges to be hired to handle a backlog of appeals."[147] The report showed that even when management was made aware of time and attendance abuse, there was little or no disciplinary action, and employees continued to cheat: "USPTO management demonstrated reluctance to take

---

[143] OPM, 2016, p. 154.

[144] OPM, 2016, p. 164.

[145] Lisa Rein, "Patent Office Filters Out Worst Telework Abuses in Report to Its Watchdog," *Washington Post*, August 10, 2014.

[146] Frederick W. Steckler, Chief Administrative Officer, "Inspector General Referral No. PPC-CI-12-1196-H, RE: Abuse of Telework Program at USPTO," memorandum for Jennifer H. Nobles, Director, Complaint Inquiry and Analysis Branch, United States Patent and Trademark Office, undated.

[147] Rein, 2014.

decisive action when the misconduct is egregious and the evidence is compelling."[148]

There were no controls in place and no efficient oversight for the USPTO telework program. Teleworking patent examiners were not required to "log into the agency's computer network or tell their supervisors the hours they worked." They did not "have to respond to a phone call from their boss the same day" it came in. The boss had "no way to tell" when employees were "at their desks," and there were little or no consequences for such work inconsistencies.[149]

The investigation uncovered the systematic abuse and inefficiency of the program and led to a joint hearing with the House Oversight and Judicial Committees convened to examine the claims.[150] Patent examiners, trademark officials, and paralegals were alleged to have abused time and attendance while teleworking. The investigation uncovered mixed results, and some claims were unproven; however, claims of end-loading—pushing work until the end of the quarter— were substantiated.[151] Additionally, claims of mortgaging—falsely claiming that work was finished during one biweekly period and then finishing the work the next period—were also substantiated.[152]

The oversight and judiciary committees provided recommendations to resolve telework issues. For example, the joint hearing recommended end-loading deterrents be put in place, along with the enforcement of work-schedule policies to prevent time and attendance fraud.[153]

---

[148] Steckler, undated, p. 9.

[149] Steckler, undated.

[150] U.S. House of Representatives, Committee on Oversight and Government Reform, "House Judiciary and Oversight Committees to Hold Joint Hearing on Patent & Trademark Office Telework Abuse," press release, November 10, 2014a.

[151] U.S. House of Representatives, Committee on Oversight and Government Reform, "Joint Hearing—Abuse of USPTO's Telework Program: Ensuring Oversight, Accountability and Quality," Washington, D.C., November 18, 2014b.

[152] U.S. House of Representatives, Committee on Oversight and Government Reform, 2014b, pp. 160–161.

[153] U.S. House of Representatives, Committee on Oversight and Government Reform, 2014b, pp. 162–163.

Additionally, the investigation suggested that USPTO should increase supervisor training, as some supervisors were not fully aware of telework policies.[154] Finally, teleworkers were directed to log into the VPN while being active on email and with online tools while they are teleworking.[155] Because the investigation was recent, we were unable to find new information on the results of the recommendations.

The experiences at USPTO demonstrate the importance of implementing effective policies, processes, training, and technologies. USPTO was not required to abolish its telework program; instead, it was required to establish the conditions for effective telework and management of teleworkers. In 2016, OPM spotlighted USPTO for its robust telework program:

> The use of telework to address limited interruptions, such as a snow day, is also part of continuity. Telework is considered a viable option especially for individuals assigned a COOP role. It is also considered for all personnel as part of the emergency preparedness plan when sufficient infrastructure is in place to allow it. . . . During the 2015 winter season, on average patent examiners maintained a nearly 92 percent production rate, and trademark examining attorneys maintained a 106 percent production rate compared to a non-inclement weather day.[156]

USPTO now has the Telework Program Office located within the office of the chief administrative officer; the Telework Program Office is responsible for setting and tracking program goals. The office collects and analyzes telework data and reports these data to leaders and managers throughout the office.[157] USPTO estimated that as of the fourth quarter of "FY 2015, the agency avoided securing $38.2 million in additional office space as a direct result of its programs."[158]

---

[154] U.S. House of Representatives, Committee on Oversight and Government Reform, 2014b, p. 167.

[155] U.S. House of Representatives, Committee on Oversight and Government Reform, 2014b, p. 164.

[156] OPM, 2016, p. 24.

[157] OPM, 2016, p. 248.

[158] OPM, 2016, p. 28.

Conclusions

We found several similarities across agencies that have telework programs, and those similarities are described in Table 3.3. Successful agency telework programs are compliant with federal and organizational policies, provide certain technological accommodations for employees who telework, and demonstrate ROI. Some of these lessons are the adaptation of performance management tools, network accommodations, and personnel training. Moreover, the analysis of USPTO exemplifies the importance of communication and performance-management metrics when implementing effective telework programs.

Table 3.3
Similarities Across Agency Telework Programs

Category	Lesson Learned
Technological	• Implement communication tools that allow teleworkers to remain involved in peer interactions. For example, NASA uses Dropbox, Google Drive, Skype, and Vidyo (a video-conferencing solution). • Ensure that remote access of data is secure with multiple encryptions (especially sensitive information), and continuously monitor and update telework technology to stay abreast of cyber threats. • Involve IT specialists in all stages of the telework planning and implementation process.
Legal	• Comply with the TEA—confirm that all employees are informed of telework eligibility.
Policy	• Require annual review of telework agreements between telework managing officers, managers, and employees to ensure that the telework program is running efficiently. • Implement a productivity-measurement tool— e.g., Asana at NASA. Data from these tools should be assessed during an employee's annual telework review. • Include telework as a part of an agency's sustainability plan. • Require recurring communication between teleworking employees and managers—e.g., daily check-ins or weekly meetings. • Analyze and update the telework program regularly through data-driven decisionmaking. • Issue agency-wide teleworking policies (should be done by upper-level management, such as the chief human capital officer).
Financial	• Create a strategic framework for identifying the value for telework investments and cost justifications to calculate ROI.

CHAPTER FOUR

Conclusions and Recommendations

Although *telework* is not consistently defined in academic and other literature, it has specific guidelines and a specific meaning in the U.S. government. The federal definition requires teleworking employees to have an approved work site rather than the ability to work from anywhere. With increasing numbers of workers desiring greater flexibility and the need for government to respond and continue operating through crisis events, it is critical that government leaders consider teleworking capabilities as a critical component of operations. Some agencies may find benefits in utilizing remote work arrangements that are even more flexible than telework is, perhaps without the requirement for an approved work site.

For agencies interested in implementing a new telework program or modifying an existing one, we found that a clear understanding of the purpose of such a program is essential for leaders who will establish the program goals, policies around different parameters, and performance measures, as well as for the managers who will be responsible for developing and implementing new technology capabilities, security protocols, and training. Based on our findings across research presented in Chapters Two and Three, we recommend the following actions for federal leaders:

- **Establish program goals that clearly explain the mission value of telework** and effectively communicate those goals to the workforce. The agencies we examined set goals to reduce real estate costs, improve employee job satisfaction, and be more responsive to the public and during crisis events.

- **Clearly communicate which job positions are eligible for telework** and which functions within each job position are suitable for off-site work. If the agency has sensitive data that require special handling, employees should be informed how to work remotely and what security protocols are required. Establishing clear policies and providing adequate training are essential to implementing the parameters that agency leaders set.
- **Create policies that document the agency's implementation of telework**, how data should be handled, and the use of government and personal computing devices. Employees and managers should have a clear understanding of whether telework is acceptable at the agency, how to effectively engage in telework, and what is expected of the teleworking employee and the teleworker's supervisor.
- **Create performance measures for the agency and teleworkers**. Agencies should measure the performance of the telework program against the established goals. For employees and managers, performance measures may consider deliverable-based or results-oriented management approaches or quantifiable metrics for performance.

Sharing best practices across federal agencies informs those with telework programs about opportunities to improve and those without telework programs about what factors need to be considered when deciding whether a telework program is appropriate for their missions. This report can serve as a tool for understanding mechanisms that can be used to accommodate changing workforces that demand flexible work hours and the option to work from alternate locations. Leaders in the IC can find additional reading on this topic in RAND's report *Moving to the Unclassified: A Guide for the Intelligence Community to Work from Unclassified Facilities*, which explores how the lessons learned in this report apply to intelligence agencies.[1]

---

[1] Weinbaum et al., 2018.

References

Allen, Tammy D., Timothy D. Golden, and Kristen M. Shockley, "How Effective Is Telecommuting? Assessing the Status of Our Scientific Findings," *Psychological Science in the Public Interest*, Vol. 16, No. 2, 2015, pp. 40–68.

Anderson, Amanda J., Seth A. Kaplan, and Ronald P. Vega, "The Impact of Telework on Emotional Experience: When, and for Whom, Does Telework Improve Daily Affective Well-Being?" *European Journal of Work and Organizational Psychology*, Vol. 24, No. 6, 2015, pp. 882–897.

Bailey, Diane E., and Nancy B. Kurland, "A Review of Telework Research: Findings, New Directions and Lessons for the Study of Modern Work," *Journal of Organizational Behavior*, Vol. 23, 2002, pp. 383–400.

Ballenstedt, Brittany, "FEMA Ramps Up Telework, Mobility," *Nextgov*, May 10, 2013. As of October 30, 2017:
http://www.nextgov.com/cio-briefing/wired-workplace/2013/05/fema-ramps-telework-mobility/63111/

Baltes, B., T. Brigges, J. Huff, J. Wright, and G. Neuman, "Flexible and Compressed Workweek Schedules: A Meta-Analysis of Their Effects on Work Related Criteria," *Journal of Applied Psychology*, Vol. 84, No. 4, 1999, pp. 496–513.

Basulto, Dominic, "The Yahoo Memo and Marissa Mayer's Big Innovation Gamble," *Washington Post*, February 28, 2013. As of November 20, 2017:
https://www.washingtonpost.com/blogs/innovations/post/the-yahoo-memo-and-marissa-mayers-big-innovation-gamble/2013/02/28/7e28266a-81b3-11e2-a671-0307392de8de_blog.html

Bayrak, Tuncay, "IT Support Services for Telecommuting Workforce," *Telematics and Informatics*, Vol. 29, No. 3, 2012, pp. 286–293.

Building Design and Construction, "House Passes 2015 GSA Budget with 17% Cut for New Construction Projects," July 23, 2014. As of June 4, 2016:
http://www.bdcnetwork.com/house-passes-2015-gsa-budget-17-cut-new-construction-projects

Bureau of Labor Statistics, U.S. Department of Labor, "24 Percent of Employed People Did Some or All of Their Work at Home in 2015," July 8, 2016. As of February 20, 2018:
https://www.bls.gov/opub/ted/2016/24-percent-of-employed-people-did-some-or-all-of-their-work-at-home-in-2015.htm

Business of Federal Technology, "FEMA and Other Agencies Showcase Successful Mobility Initiatives," April 30, 2013. As of February 20, 2017:
https://fcw.com/articles/2013/04/30/fema-telework.aspx

Caillier, James G., "The Impact of Teleworking on Work Motivation in a U.S. Federal Government Agency," *American Review of Public Administration*, Vol. 42, No. 4, 2012, pp. 461–480.

————, "Are Teleworkers Less Likely to Report Leave Intentions in the United States Federal Government Than Non-Teleworkers Are?" *American Review of Public Administration*, Vol. 43, No. 1, 2013a, pp. 72–88.

————, "Satisfaction with Work-Life Benefits and Organizational Commitment/Job Involvement: Is There a Connection?" *Review of Public Personnel Administration*, Vol. 33, No. 4, 2013b, pp. 340–364.

Choo, S., P. Mokhtarian, and I. Salomon, "Does Telecommuting Reduce Vehicle Miles Traveled? An Aggregate Time Series Analysis for the U.S.," *Transportation*, Vol. 32, No. 1, 2005, pp. 37–64.

Coleman, Casey, "Total Workplace Transforms Federal Office Space," *Around the Corner*, November 19, 2013. As of February 20, 2017:
https://gsablogs.gsa.gov/innovation/2013/11/19/gsas-total-workplace-creates-a-21st-century-workplace-designed-to-save-money-and-increase-efficiency-and-productivity/

Computer Connections/Metropolitan Washington Council of Governments, *Employer Telework Case Study*, Washington, D.C., 2010. As of October 30, 2017:
http://www.federaletc.org/pdf/EmployerTeleworkCaseStudy-NRC.pdf

Dahlstrom, Timothy R., "Telecommuting and Leadership Style," *Public Personnel Management*, Vol. 42, No. 3, 2013, pp. 438–451.

Daly, Jimmy, "FEMA Is Ready for Anything—Even Telework," *FedTech*, May 24, 2013. As of February 20, 2017:
http://www.fedtechmagazine.com/article/2013/05/fema-ready-anything-even-telework

Federal Emergency Management Agency, "Continuity Planning for Telework," Washington, D.C., undated-a. As of October 11, 2017:
http://www.fema.gov/media-library-data/1410875825235-6e7eccb0ed3d181fda03dcb780626e18/COOP%20Telework%20Planning.pdf

————, "Lesson 2: Elements of a Viable COOP," in *Continuity of Operations Planning Web-Based Course*, Washington, D.C., undated-b, pp. 1–7. As of June 1, 2016:
https://training.fema.gov/hiedu/docs/cgo/
week%204%20-%20lesson%202%20-%20elements%20of%20a%20viable%20
coop.pdf

————, *FEMA Manual 3000.3*, Washington, D.C., 2000. As of October 11, 2017:
http://www.fema.gov/pdf/library/3000_3.pdf

————, *Department and Agency: Continuity Telework Exercise; Exercise Plan (EXPLAN) Template*, Washington, D.C., May 2013. As of October 11, 2017:
https://www.fema.gov/media-library/assets/documents/86274

————, "About the Agency," webpage, last updated May 16, 2017. As of October 11, 2017:
https://www.fema.gov/about-agency

FEMA—*See* Federal Emergency Management Agency.

Fonner, Kathryn L., and Michael Roloff, "Why Teleworkers Are More Satisfied with Their Jobs Than Are Office-Based Workers: When Less Contact Is Beneficial," *Journal of Applied Communication Research*, Vol. 38, No. 4, 2010, pp. 336–361.

Friedman, Stewart D., and Alyssa Westring, "Empowering Individuals to Integrate Work and Life: Insights for Management Development," *Journal of Management Development*, Vol. 34, No. 3, 2015, pp. 299–315.

GAO—*See* U.S. Government Accountability Office.

General Services Administration, *GSA Mobility and Telework Policy*, Washington, D.C., undated. As of March 21, 2018:
https://www.gsa.gov/cdnstatic/GSAteleworkpolicy.pdf

————, *GSA Telework Program Management Office: Recipe Book*, Washington, D.C., 2011.

————, *General Services Administration (GSA) Information Technology (IT) Standards Profile*, GSA Order, CIO P 2160.1E, Washington, D.C., February 2014. As of October 30, 2017:
https://www.gsa.gov/cdnstatic/CIO_P_2160.1E_General_Services_
Administration_%28GSA%29_Information_Technology_%28IT%29_
Standards_Profile_%28Final_2-12-2014%29_Rv_May_9__2016.pdf

————, "Background and History," webpage, last reviewed August 13, 2017a. As of February 20, 2017:
https://www.gsa.gov/portal/category/21354

————, "Regional Buildings Overview," webpage, last reviewed August 13, 2017b. As of October 30, 2017:
http://www.gsa.gov/portal/category/21660

———, "Resources for Managing Teleworkers," webpage, last reviewed August 13, 2017c. As of October 30, 2017:
https://www.gsa.gov/portal/category/102107

———, "Telework Dashboards," webpage, last reviewed August 13, 2017d. As October 30, 2017:
http://www.gsa.gov/portal/content/114495

———, *Telework Training*, Washington, D.C., last reviewed August 13, 2017e.
https://www.gsa.gov/portal/category/102551

———, "The Goals of Telework," webpage, last updated August 28, 2017f. As of November 21, 2017:
https://www.gsa.gov/portal/content/114487

General Services Administration, Office of Inspector General, *GSA's Program for Managing Virtual Employees and Teleworkers Needs Improvement*, Washington, D.C., 2015a. As of October 11, 2017:
https://gsaig.gov/content/gsa%E2%80%99s-program-managing-virtual-employees-and-teleworkers-needs-improvement-a130019-1

———, *Status of Recommendations: Audit of NRC's Full-Time Telework Program*, Washington, D.C., 2015b, pp. 1–8.

Georgetown University Law Center, *Telework in the Federal Government: The Overview Memo*, Washington, D.C., 2009. As of October 11, 2017:
http://scholarship.law.georgetown.edu/cgi/
viewcontent.cgi?article=1015&context=legal

GSA—*See* General Services Administration.

Hammer, Leslie, M. Neal, J. Newsom, K. Brockwood, and C. Colton, "A Longitudinal Study of the Effects of Dual-Earner Couples' Utilization of Family-Friendly Workplace Supports on Work and Family Outcomes," *Journal of Applied Psychology*, Vol. 90, No. 4, 2005, pp. 799–810.

Internal Revenue Service, "At a Glance: IRS Divisions and Principal Offices," webpage, last reviewed July 27, 2017a. As of October 30, 2017:
https://www.irs.gov/uac/at-a-glance-irs-divisions-and-principal-offices

———, "The Agency, Its Mission and Statutory Authority," webpage, last reviewed August 6, 2017b. As of October 30, 2017:
https://www.irs.gov/uac/the-agency-its-mission-and-statutory-authority

———, "Part 6: Human Resources Management, Chapter 800. 0 Employee Benefits, Section 2: IRS Telework Program," in *Internal Revenue Manuals*, Washington, D.C., last reviewed September 10, 2017c. As of October 30, 2017:
https://www.irs.gov/irm/part6/irm_06-800-002.html#d0e10

Internal Revenue Service and National Treasury Employee Union, *2012 National Agreement II*, Washington, D.C., 2012.

IRS—*See* Internal Revenue Service.

Jones, Jeffrey M., "In U.S., Telecommuting for Work Climbs to 37%," *Gallup News*, August 19, 2015. As of May 20, 2016:
http://www.gallup.com/poll/184649/telecommuting-work-climbs.aspx

Kopp, Emily, "Don't Ask GSA How Many Full-Time Teleworkers It Has—It Doesn't Know," *Federal News Radio*, January 21, 2015. As of February 20, 2017:
https://federalnewsradio.com/management/2015/01/
dont-ask-gsa-how-many-full-time-teleworkers-it-has-it-doesnt-know/

Kurland, Nancy B., and Terri D. Egan, "Telecommuting: Justice and Control in the Virtual Organization," *Organization Science*, Vol. 10, No. 4, 1999, pp. 500–513.

Lautsch, B. A., E. E. Kossek, and S. C. Eaton, "Supervisory Approaches and Paradoxes in Managing Telecommuting Implementation," *Human Relations*, Vol. 62, No. 6, 2009, pp. 795–827.

Lister, Kate, "Latest Telecommuting Statistics," webpage, GlobalWorkplaceAnalytics.com, January 2016. As of May 20, 2016:
http://globalworkplaceanalytics.com/telecommuting-statistics

Lleywellyn, Ali, "Teleworking," *Open NASA*, November 15, 2011. As of March 21, 2018:
http://www.opennasa.org/teleworking.html

Mazid, Rami, "Cisco Study Finds Telecommuting Significantly Increases Employee Productivity, Work-Life Flexibility and Job Satisfaction," Cisco, June 26, 2009. As of February 20, 2016:
https://newsroom.cisco.com/press-release-content?articleId=5000107

McLaughlin, Matt, "Telework Thrives as Mobility Grows," *FedTech*, April 17, 2012. As of February 20, 2017:
http://www.fedtechmagazine.com/article/2012/04/telework-thrives-mobility-grows

Mobile Work Exchange, *The Telework Revolution: Bringing Theory to Practice*, Alexandria, Va., 2013.

Mokhtarian, Patricia L., "Defining Telecommuting," *Transportation Research Record*, Vol. 1305, 1991, pp. 273–281.

Mokhtarian, Patricia L., and Ilan Salomon, "Modeling the Desire to Telecommute: The Importance of Attitudinal Factors in Behavioral Models," *Transportation Research Part A: Policy and Practice*, Vol. 31, No. 1, 1997, pp. 35–50.

Morganson, V. J., D. A. Major, K. L. Oborn, J. M. Verive, and M. P. Heelan, "Comparing Telework Locations and Traditional Work Arrangements: Differences in Work-Life Balance Support, Job Satisfaction, and Inclusion," *Journal of Managerial Psychology*, Vol. 25, No. 6, 2010, pp. 578–595.

Moss, Hannah, "What Your Agency Can Learn from NASA," *Gov Loop*, December 11, 2014. As of February 20, 2017:
https://www.govloop.com/nasa-at-your-agency/

NASA—*See* National Aeronautics and Space Administration.

National Aeronautics and Space Administration, "Remote Access Services," webpage, Information Technology and Communication Division, undated-a. As of October 12, 2017:
https://www.hq.nasa.gov/office/itcd/remote_access.html

———, "Work from Anywhere: Factsheet," undated-b. As of February 20, 2017:
http://www.hq.nasa.gov/office/hqhr/docs/Factsheet.pdf

———, *NASA Desk Guide on Telework Programs*, Washington, D.C.: Office of Human Capital Management, NSREF-3000-0012, April 2010a. As of October 12, 2017:
https://searchpub.nssc.nasa.gov/servlet/sm.web.Fetch/
Telework_Desk_Guide.pdf?rhid=1000&did=778067&type=released

———, *NASA Procedural Requirements*, Washington, D.C.: Office of Human Capital Management, NPR 3600.2, 2010b. As of October 12, 2017:
http://nodis3.gsfc.nasa.gov/npg_img/N_PR_3600_0002_/
N_PR_3600_0002_.pdf

———, *Work from Anywhere: NASA's Telework Program*, Stennis Space Center, Miss.: NASA Shared Services Center, March 2013. As of October 12, 2017:
https://searchpub.nssc.nasa.gov/servlet/sm.web.Fetch/
2013_Telework_-_Final.pdf?rhid=1000&did=1501938&type=released

———, *FY 2015 Agency Financial Report*, Washington, D.C., 2015. As of October 12, 2017:
http://www.nasa.gov/sites/default/files/atoms/files/fy2015_afr_11-16-15.pdf

National Science Foundation, "About the National Science Foundation," webpage, undated. As of May 29, 2016:
http://www.nsf.gov/about/

———, "'Telework' Benefits Employers, Employees and the Environment," news release, March 11, 2008a. As of October 12, 2017:
https://www.nsf.gov/news/news_summ.jsp?cntn_id=111252

———, *Telework Under the Microscope: A Report on the National Science Foundation's Telework Program*, Alexandria, Va., 2008b.

———, *FY 2015 Strategic Sustainability Performance Plan*, Alexandria, Va., 2015. As of November 16, 2017:
https://www.nsf.gov/pubs/2016/nsf16025/nsf16025.pdf

Nilles, Jack M., *Telecommuting Forecasts*, Los Angeles, Calif.: Telecommuting Research Institute, 1991.

Nilles, Jack M., F. Roy Carlson, Jr., Paul Gray, and Gerhard J. Hanneman, *The Telecommunications-Transportation Tradeoff: Options for Tomorrow*, New York: John Wiley and Sons, 1976.

NRC—*See* Nuclear Regulatory Commission.

NSF— *See* National Science Foundation.

Nuclear Regulatory Commission, "About NRC," webpage, undated. As of June 3, 2016:
http://www.nrc.gov/about-nrc.html

———, *Audit Report: Audit of NRC's Full-Time Telework Program*, Rockville, Md., OIG-14-A-05, December 2013. As of October 12, 2017:
http://www.nrc.gov/docs/ML1334/ML13345A194.pdf

———, "Locations," webpage, last reviewed September 25, 2017. As of February 20, 2017:
https://www.nrc.gov/about-nrc/locations.html

OPM—*See* U.S. Office of Personnel Management.

Park, George S., Jack M. Baer, and Walter S. Nilles, *Trends and Factors Influencing Telecommuting in Southern California*, Santa Monica, Calif.: RAND Corporation, DRU-1465-SCTP, 1996. As of October 12, 2017:
https://www.rand.org/pubs/drafts/DRU1465.html

Peters, P., K. Tijdens, and C. Wetzels, "Employees' Opportunities, Preferences, and Practices in Telecommuting Adoption," *Information and Management*, Vol. 41, No. 4, 2004, pp. 469–482.

Pinsonneault, A., *The Impacts of Telecommuting on Organizations and Individuals: A Review of the Literature*, Montreal: HEC Montreal, 1999.

Popuri, Y. D., and C. R. Bhat, "On Modeling Choice and Frequency of Home-Based Telecommuting," *Transportation Research Record*, Vol. 1858, 2003, pp. 55–60.

Public Law 93-579, Privacy Act of 1974, December 31, 1974.

Public Law 111-5, American Recovery and Reinvestment Act of 2009, February 17, 2009.

Public Law 111-292, The Telework Enhancement Act of 2010, December 9, 2010.

Ravindranath, Mohana, "Is Telework a Growing Cyber Threat? New Guidelines Offer Security Tips," *Nextgov*, March 16, 2016. As of October 12, 2017:
http://www.nextgov.com/cybersecurity/2016/03/
telework-leading-more-data-breaches-new-guidelines-offer-security-tips/126732/

Reaney, Patricia, "About 20 Percent of Global Workers Telecommute: Poll," Reuters, January 24, 2012. As of October 12, 2017: http://www.huffingtonpost.com/2012/01/24/ workers-telecommute_n_1228004.html

Rein, Lisa, "Patent Office Filters Out Worst Telework Abuses in Report to Its Watchdog," *Washington Post*, August 10, 2014. As of February 27, 2017: https://www.washingtonpost.com/politics/ patent-office-filters-out-worst-telework-abuses-in-report-to-watchdog/2014/08/10/ cd5f442e-1e4d-11e4-82f9-2cd6fa8da5c4_story.html?utm_term=.4a61a436bef5

Ruth, Stephen, and Imran Chaudhry, "Telework: A Productivity Paradox?" *IEEE Internet Computing*, Vol. 12, No. 6, 2008, pp. 87–90.

Schectman, Joel, "GSA Enables More Telework," *Wall Street Journal*, September 20, 2013. As of October 13, 2017: http://blogs.wsj.com/cio/2013/09/20/gsa-enables-more-telework/

Shive, David, "Federal Workforce on Duty During Winter Storm Jonas," *Around the Corner*, General Services Administration, January 28, 2016. As of February 20, 2017: https://gsablogs.gsa.gov/innovation/2016/01/28/ federal-workforce-on-duty-during-winter-storm-jonas/

Steckler, Frederick W., Chief Administrative Officer, "Inspector General Referral No. PPC-CI-12-1196-H, RE: Abuse of Telework Program at USPTO," memorandum for Jennifer H. Nobles, Director, Complaint Inquiry and Analysis Branch, United States Patent and Trademark Office, undated. As of February 27, 2017: http://apps.washingtonpost.com/g/page/politics/initial-report-on-us-patent-and-trademark-office-investigation-of-telework-fraud-allegations/1244/

Sullivan, J., "How Yahoo's Decision to Stop Telecommuting Will Increase Innovation," ERE Media, February 26, 2013. As of October 13, 2017: http://www.eremedia.com/ere/ how-yahoos-decision-to-stop-telecommuting-will-increase-innovation/

Taskin, L., and F. Bridoux, "Telework: A Challenge to Knowledge Transfer in Organizations," *International Journal of Human Resource Management*, Vol. 21, No. 13, 2010, pp. 2503–2520.

Telework.gov, "Telework Enhancement Act," webpage, undated. As of October 13, 2017: https://www.telework.gov/guidance-legislation/telework-legislation/ telework-enhancement-act/

TIGTA—*See* Treasury Inspector General for Tax Administration.

Treasury Inspector General for Tax Administration, *Significant Additional Real Estate Cost Savings Can Be Achieved by Implementing a Telework Workstation Sharing Strategy*, Washington, D.C.: U.S. Department of the Treasury, August 27, 2012a. As of November 26, 2017:
https://www.treasury.gov/tigta/auditreports/2012reports/201210100fr.pdf

———, "TIGTA: IRS Can Reduce Real-Estate Costs by Increasing Use of Telework," press release, September 24, 2012b. As of October 13, 2017:
https://www.treasury.gov/tigta/press/press_tigta-2012-47.htm

Treasury Inspector General for Tax Administration, Office of Inspections and Evaluations, *Review of the Implementation of the Telework Enhancement Act of 2010*, Washington, D.C., U.S. Department of the Treasury, July 2013. As of October 13, 2017:
https://www.treasury.gov/tigta/iereports/2013reports/2013IER006fr.pdf

———, *Telework Qualification Requirements Are Generally Being Met, but Program Improvements Are Needed*, Washington, D.C.: U.S. Department of the Treasury, July 19, 2016. As of February 20, 2018:
https://www.treasury.gov/tigta/auditreports/2016reports/201610039fr.pdf

Tugend, Alina, "It's Unclearly Defined, but Telecommuting Is Fast on the Rise," *New York Times*, March 7, 2014. As of May 21, 2016:
http://www.nytimes.com/2014/03/08/your-money/when-working-in-your-pajamas-is-more-productive.html?_r=0

United States Patent and Trademark Office, "About Us," webpage, undated. As of October 13, 2017:
https://www.uspto.gov/about-us

U.S. Department of Transportation, *Transportation Implications of Telecommuting*, Washington, D.C.: Research and Innovation Technology Administration, 1993. As of May 21, 2016:
http://ntl.bts.gov/DOCS/telecommute.html#TOP

U.S. Government Accountability Office, *Federal Telework: Better Guidance Could Help Agencies Calculate Benefits and Costs*, Washington, D.C., GAO-16-551, July 2016.

U.S. House of Representatives, Committee on Oversight and Government Reform, "House Judiciary and Oversight Committees to Hold Joint Hearing on Patent & Trademark Office Telework Abuse," press release, November 10, 2014a. As of February 27, 2017:
https://oversight.house.gov/release/house-judiciary-oversight-committees-hold-joint-hearing-patent-trademark-office-telework-abuse/

————, "Joint Hearing—Abuse of USPTO's Telework Program: Ensuring Oversight, Accountability and Quality," Washington, D.C., November 18, 2014b. As of October 12, 2017:
https://oversight.house.gov/hearing/joint-hearing-abuse-usptos-telework-program-ensuring-oversight-accountability-quality/

U.S. Office of Personnel Management, *Guide to Telework in the Federal Government*, Washington, D.C., April 2011. As of October 13, 2017:
https://www.telework.gov/guidance-legislation/telework-guidance/telework-guide/guide-to-telework-in-the-federal-government.pdf

————, *2014 Status of Telework in the Government Report to Congress: Fiscal Year 2013*, Washington, D.C., November 2015. As of October 13, 2017:
https://www.telework.gov/reports-studies/reports-to-congress/2014-report-to-congress.pdf

————, *Status of Telework in the Federal Government: Report to Congress; Fiscal Years 2014–2015*, Washington, D.C., November 2016. As of October 13, 2017:
https://www.telework.gov/reports-studies/reports-to-congress/2016-report-to-congress.pdf

USPTO—*See* United States Patent and Trademark Office.

Vega, Ronald P., Amanda J. Anderson, and Seth A. Kaplan, "A Within-Person Examination of the Effects of Telework," *Journal of Business and Psychology*, Vol. 30, No. 2, 2015, pp. 313–323.

Weinbaum, Cortney, Arthur Chan, Karlyn D. Stanley, and Abby Schendt, *Moving to the Unclassified: How the Intelligence Community Can Work from Unclassified Facilities*, Santa Monica, Calif.: RAND Corporation, RR-2024-OSD, 2018. As of April 30, 2018:
http://www.rand.org/pubs/research_reports/RR2024.html

Weinbaum, Cortney, Richard Girven, and Jenny Oberholtzer, *The Millennial Generation: Implications for the Intelligence and Policy Communities*, Santa Monica, Calif.: RAND Corporation, RR-1306-OSD, 2016. As of October 13, 2017:
http://www.rand.org/pubs/research_reports/RR1306.html

Westfall, Ralph D., "Does Telecommuting Really Increase Productivity? Fifteen Rival Hypotheses," *AMCIS 1997 Proceedings*, Association for Information Systems Electronic Library, 1997.

Yu, S., "How to Make Teleworking Work," *Communications News*, Vol. 45, No. 12, 2008, pp. 30–32.

Zhu, P., "Are Telecommuting and Personal Travel Compliments or Substitutes?" *Annals of Regional Science*, Vol. 48, No. 2, 2012, pp. 619–639.